T0141988

Evolution of Semantic Systems

Bernd-Olaf Küppers • Udo Hahn
Stefan Artmann

Editors

Evolution of Semantic Systems

Springer

Editors
Bernd-Olaf Küppers
Frege Centre for Structural Sciences
Friedrich Schiller University Jena
Jena
Germany

Stefan Artmann
Frege Centre for Structural Sciences
Friedrich Schiller University Jena
Jena
Germany

Udo Hahn
Frege Centre for Structural Sciences, and
Jena University Language & Information
    Engineering (JULIE) Lab
Friedrich Schiller University Jena
Jena
Germany

ISBN 978-3-642-43442-6   ISBN 978-3-642-34997-3 (eBook)
DOI 10.1007/978-3-642-34997-3
Springer Heidelberg New York Dordrecht London

ACM Computing Classification (1998): I.2, H.1, E.4

Printed on acid-free paper

Springer is part of Springer Science+Business Media (www.springer.com)

Preface

Complex systems in nature and society make use of information for the development of their internal organisation and the control of their functional mechanisms. Alongside technical aspects of storing, transmitting and processing information, the various semantic aspects of information—such as meaning, sense, reference and function—play a decisive part in the analysis of such systems. This raises important questions:

- What are paradigmatic examples of semantic systems—formal languages, natural languages, programming languages and corresponding coding artefacts (programs), ontologies, the genetic code, neural nets, animal signal systems, the World Wide Web or social organisations?
- Are there universal principles underlying the origin and evolution of semantic systems?
- Can we draw a rigid line of demarcation between semantic and non-semantic systems, or is the transition between the two types of systems a smooth one?
- Does the understanding of semantic systems necessarily imply a "naturalisation" of semantics?

A large variety of scientific disciplines are required to cooperate if adequate answers to these and further related questions are to be developed. With the aim of fostering a better understanding of semantic systems from an evolutionary and multidisciplinary perspective, this volume collates contributions by philosophers and natural scientists, linguists and information and computer scientists. They do not follow a single research paradigm—rather, they shed, in a complementary way, new light upon some of the most important aspects of the evolution of semantic systems. The eleven chapters of this volume are arranged according to an increasing level of specificity of their respective topics.

The contributions of Dagfinn Føllesdal and Stefan Artmann are concerned with philosophical questions that arise when fundamental concepts of semantics, such as meaning and reference, become the object of analytical and pragmatist philosophy.

In "The Emergence of Reference", Føllesdal asks which properties objects must possess so that they can be referred to. He presents three such features. First of all, objects must have numerous properties that, most frequently, are only partially known by us but which can be explored. Secondly, objects must be temporal entities that retain their identity when changing their properties. Thirdly, while we may falsely ascribe properties to an object, it must be possible that our false ascriptions concern the same object as do our corrected, true ascriptions. Føllesdal shows how these three features give objects such great epistemological importance for human cognition and how they enable language to keep track of objects.

Artmann's paper on "Pragmatism and the Evolution of Semantic Systems" also investigates reference, yet it takes its point of departure in pragmatist philosophy, namely in Charles S. Peirce's evolutionary pragmatism and its naturalistic reconstruction by John Dewey. How do these philosophers understand the origin and development of meaning, and how might their theories contribute to an explanation of the evolution of semantic systems? Artmann maintains that pragmatism could play a very important role in our understanding of the dynamics of meaning if it were to integrate theories and models from structural science. His analyses of classical pragmatism lead to the project of developing a modern version of pragmatist thought, which he calls "structural pragmatism".

After Føllesdal's and Artmann's philosophical papers, the next three chapters discuss abstract structural-scientific concepts that are necessary for the development of an evolutionary theory of semantic systems: the concepts of system, semantics and structure.

Systems, in general, are the topic of Klaus Kornwachs' paper on "System Surfaces—There Is Never Just Only One Structure". Kornwachs focuses on the concept of surface of a system: does it refer to the top layer of a real object or to the construction of the mind that is useful when the world is conceived of in terms of systems? Kornwachs tackles this question by looking into the history of systems theory and reconstructing the relationships that philosophers and scientists have postulated between the ideas of system and order, on the one hand, and between the ideas of system and process, on the other. How system theorists think about those relationships will have a great influence on how they conceptualise the evolution of semantic systems.

In "Elements of a Semantic Code", Bernd-Olaf Küppers presents a new approach to the objectification and quantification of meaning-bearing information. His theory of semantic code addresses general aspects of semantics, such as novelty, pragmatic relevance, selectivity and complexity. He integrates these into a reservoir of value elements from which a recipient configures a value scale for the particular evaluation of information. Küppers' approach to a quantified theory of semantic information is a new example of how structural sciences—such as cybernetics, information theory and systems theory—proceed when analysing complex systems in nature and society: the comprehensive, abstract structures that they describe are not restricted to a particular class of complex systems, such as natural or artificial, living or non-living ones.

Both Kornwachs' theory of systems and Küppers' theory of semantics follow a structural-scientific approach to the evolution of semantic systems. In doing so, they rely heavily upon the concept of structure. In "Talking about Structures", Rüdiger Inhetveen and Bernhard Schiemann show how that concept has become, during the second half of the twentieth century, an interdisciplinary point of reference within a wide range of scientific disciplines transcending the frontiers between formal sciences, empirical sciences and humanities.

The four subsequent contributions analyse problems in the evolution of semantic systems from the perspectives of applied structural sciences: information science (Jerry R. Hobbs and Rutu Mulkar-Mehta), artificial intelligence (Jon Doyle), computational linguistics (Udo Hahn) and computer science (Ian Horrocks).

Hobbs and Mulkar-Mehta's "Toward a Theory of Information Structure" presents a formal theory of the structure of information that, on the one hand, is not restricted to a particular type of medium in which information is made concrete, and, on the other hand, is based on cognitive foundations of the use of semantic systems by human beings. Hobbs and Mulkar-Mehta proceed by discussing questions of compositionality and co-reference and by giving some applications of their theory of information structure to the translation of natural language into logical form as well as to the analysis of diagrams and Gantt charts.

In "Mechanics and Mental Change", Doyle examines theories of mental change with the aim of assessing their adequacy for characterising realistic limits on cognitive processes. His motivation to do so is that, in real life, human rationality departs from ideal theories of rationality and meaning as developed in epistemology and economics. In the real world, human beings do not learn and deliberate effortlessly, nor are they deterred by ignorance and inconsistency from taking action. By using a simple reasoning system taken from artificial intelligence Doyle shows how mechanical concepts—such as inertia, force, work and constitutional elasticity—can be applied to the analysis of limits on cognitive systems.

Hahn's paper on "Semantic Technologies: A Computational Paradigm for Making Sense of Informal Meaning Structures" addresses a problem that is also of great interest in the field of artificial intelligence. Although the World Wide Web has become the largest knowledge repository ever available, the present-day computational infrastructure cannot cope adequately with the meaning captured in its interlinked documents. The new paradigm of semantic technologies might help unlock this information with computational means. It is characterised by a focus on qualitative, un(der)structured data and aims to understand the content of Web documents. In his paper, Hahn introduces fundamental building blocks of semantic technologies, namely terms, relationships between terms and formal means to reason over relationships. He also elaborates on the particular conditions imposed by the Web (e.g., massive amounts of "noisy" data characterised by incompleteness, redundancy, inconsistency, different degrees of granularity and trustworthiness) that tend to impose entirely new challenges on (logic-based) formal approaches to semantic information processing, and he focuses on the crucial role that ontologies play in "semanticising" the Web.

Taking up Hahn's perspective on the importance of ontologies as a central methodology of semantic technologies, Horrocks' contribution is provocatively entitled "What Are Ontologies Good For"? He defines ontologies as an engineering artefact that is usually a conceptual model of some slice of the real world and typically formalised in terms of a suitable logic that makes it possible to apply automated reasoning in ontology design and deployment. As Horrocks emphasises, the tremendous value of this semantic technology has already been demonstrated in numerous medical applications. But he also points out that it is still a difficult and resource-consuming task to develop and maintain ontologies of good quality, so that the deployment of ontology-based information systems appears likely to be quite limited.

The last two chapters of this compilation show how scientific semantic change can be analysed from the perspective of information science and philosophy of science.

Alexa McCray and Kyungjoon Lee report, in "Taxonomic Change as a Reflection of Progress in a Scientific Discipline", on their study of the evolution of the "Medical Subject Headings" (MeSH) over the last 45 years. MeSH is a terminological system that is used to index a large proportion of the world's biomedical literature for information-retrieval purposes in the context of MEDLINE, the U.S. National Library of Medicine's bibliographic retrieval system. McCray and Lee show that the evolution of MeSH as a semantic system is shaped by internal factors (e.g., the need for developing a more fundamental ontological structure) and external factors (e.g., the development of biomedical knowledge). From case studies in the evolution of semantic systems, such as MeSH, important conclusions for research in ontology evolution and for the conceptual modelling and evolution of knowledge domains can be drawn.

Finally, C. Ulises Moulines' paper on "Crystallization as a Form of Scientific Semantic Change: The Case of Thermodynamics" reconstructs another evolutionary process of scientific knowledge. He presents a classification of four ideal types of change in the semantic systems that constitute scientific disciplines: normal evolution, revolutionary replacement, embedding of a conceptual system into another one and gradual crystallisation of a conceptual framework. Moulines focuses on the last type of change, which comprises a piecemeal, yet fundamental change in the semantic system of a discipline over a relatively long period of time. His case study reconstructs, within the framework of the structuralistic philosophy of science, the gradual emergence of phenomenological thermodynamics in the middle of the nineteenth century.

The papers of Doyle, Føllesdal, Hobbs and Mehta, Horrocks, Inhetveen and Schiemann, Küppers, McCray and Lee were presented at the symposium *Evolution of Semantic Systems* (Jena, Germany; October 5th and 6th, 2007). This workshop was organised by three institutions of the Friedrich Schiller University at Jena: the Institute of Philosophy, the Jena University Language and Information Engineering (JULIE) Laboratory and the Institute of Computer Science. The editors would like to thank the speakers (who also included Christoph von der Malsburg, Barry Smith and Nicola Guarino) who contributed to the success of that symposium, as well as their colleagues Clemens Beckstein, Peter Dittrich and Harald Sack for their help in its organisation. Thanks also go to Sara Neuhauser and Gudrun Schmidt for preparing and reformatting the submitted contributions for publication.

Jena, Germany   Bernd-Olaf Küppers, Udo Hahn, Stefan Artmann
August 2012

Contents

List of Contributors

Stefan Artmann Frege Centre for Structural Sciences, Friedrich Schiller University Jena, Jena, Germany

Jon Doyle Department of Computer Science, North Carolina State University, Raleigh, NC, USA

Dagfinn Føllesdal Department of Philosophy, Stanford University, and Center for the Study of Mind in Nature, University of Oslo, 0316 Oslo, Norway

Udo Hahn Department of German Linguistics, Friedrich Schiller University Jena, Germany

Jerry R. Hobbs Information Sciences Institute, University of Southern Caifornia, Admiralty Way, Marina del Rey, CA, USA

Ian Horrocks Oxford University, Department of Computer Science, Oxford, UK

Rüdiger Inhetveen Former Chair for Artificial Intelligence and Center for Ethics and Scientific Communication, Friedrich-Alexander-Universität Erlangen-Nürnberg Erlangen, Germany

Klaus Kornwachs Brandenburgische Technische Universität Cottbus, Cottbus, Germany

Bernd-Olaf Küppers Frege-Centre for Structural Sciences, Friedrich Schiller University Jena, Jena, Germany

Kyungjoon Lee Center for Biomedical Informatics, Harvard Medical School, Boston, MA, USA

Alexa T. McCray Center for Biomedical Informatics, Harvard Medical School, Boston, MA, USA

C. Ulises Moulines Department of Philosophy, Logic and Philosophy of Science, Ludwig-Maximilians-Universität München, Munich, Germany

Rutu Mulkar-Mehta Precyse Advanced Technologies, Founders Parkway, Suite, Alpharetta, GA

Bernhard Schiemann Former Chair for Artificial Intelligence and Center for Ethics and Scientific Communication, Friedrich-Alexander-Universität Erlangen-Nürnberg Erlangen, Germany

The Emergence of Reference

Dagfinn Føllesdal

Abstract The central theme of this volume, the evolution of semantic systems, raises issues about the general idea of a system, what differentiates semantic systems from other systems, and about the two key topics of semantics: meaning and reference. The paper focuses on the emergence of reference and how this is connected with individuation, reification and the notion of an object. Objects have three features that are crucial for reference: First: They are the bearers of a (usually) large number of properties and relations. Normally we know only a small number of these, but the object is conceived of as having numerous further properties that we do not know yet, but which are there to be explored. Secondly: Objects, except mathematical ones and a few others, change over time. One and the same object can have a property at one time and lack it at another time. The object remains identical through changes. Modalities come in at this point; not only are there the actual changes, but there are also possible ones, there are accidents and there are necessities. Or, at least, so we say. Finally: There is our fallibility. We may have false beliefs about objects. We may seek to correct these beliefs, but all the while our beliefs, true or false, are of the objects in question. A belief, or set of beliefs, is not about whichever object happens best to satisfy our beliefs ([1]). A semantics that just would seek to maximize our set of true beliefs would reflect poorly the role that objects play in epistemology. These three important features of objects are reflected in the way names and other referring expressions function in language. Objects are crucially important for our daily lives, our cognition and our semantic theory, and keeping track of objects is an important feature of language. How this happens is a major theme of the paper.

D. Føllesdal (✉)
Department of Philosophy, Stanford University, and Center for the Study of Mind in Nature, University of Oslo, 0316 Oslo, Norway
e-mail: dagfinn@csli.stanford.edu

B.-O. Küppers et al. (eds.), *Evolution of Semantic Systems*,
DOI 10.1007/978-3-642-34997-3_1, © Springer-Verlag Berlin Heidelberg 2013

1 Introduction

Among the questions that have been discussed at the conference I will concentrate on two and for each of them address a number of the sub-questions that have been indicated:

Question 1:   *Semantic systems*—What are paradigmatic examples of semantic systems: formal languages, natural languages, programming languages, ontologies, the genetic code, neural nets, animal signal systems, computer systems and social systems?

Question 2:   *Semantic versus non-semantic systems*—Does there exist a rigid line of demarcation between *semantic* and *non-semantic* systems, or is the transition between both types of systems a smooth one?

2 Semantic Systems

What is a system? As a first approximation, which may be too broad, but suffices for our purposes, we may define a system as a *structure*, that is, *set of entities* (nodes) with *relations* defined between them. What then are semantic systems?

In traditional linguistics it is customary to distinguish between *syntax, phonology* and *semantics*. Are these three disciplines *autonomous* or do they depend upon one another, if so how?

Syntax depends on semantics: syntax studies those structural features that are relevant to meaningfulness. Likewise, *semantics* is dependent upon syntax: while syntax focuses on those substructures, or *expressions*, that occur in communicative activity and studies their structure, semantics is supposed to explain how these expressions that occur in normal speech and writing have a meaning, how they make sense and function in communication and what meaning they have. *Phonology* explores differences in sound where they matter for semantics.

3 Semantic Versus Non-semantic Systems

Now on to the second main question: Does there exist a rigid line of demarcation between *semantic* and *non-semantic* systems, or is the transition between both types of systems a smooth one?

My tentative answer to this question is: a system becomes a semantic system when it becomes endowed with meaning. Let us see how.

3.1 Semantics

Structures can be used to communicate. Instead of studying the structure and talking about it, we use the structure to communicate. What makes a structure a language is that it is used to communicate. Semantics is a study of what the communication consists in and how the various features of the structure contribute to the communication (words, arrangement of words into more complex units, like sentences, etc.)

3.2 Slide Rule as an Example

Slide rules are now obsolete; students nowadays have barely heard of it and do not know how it functions.

A slide rule is a simple system, two rods with numerical scales that slide against one another. Through the use we make of it it becomes a simple semantic system. This semantic system incorporates a mathematical structure: instead of multiplying numbers one adds the corresponding exponents. The rods and their movements are given meaning, which makes them part of a semantic system that also includes some mathematics.

The rods could be given other meanings; one could use them for simple addition and also for many other purposes. Not the physical system itself is semantic, but the system *under an interpretation*.

3.3 Some Small Points About Semantic Systems

1. The structure by itself is not a semantic system, it becomes a semantic system through interpretation, and one and the same structure can become many different semantic systems.
2. What makes it semantic is the interpretation: its *use* in *publicly observable* ways. This is the first step in answering the "how" question: How does a system become a semantic system? This is only a first small step; lots of further details are needed.

   Are *neural nets* semantic systems?

   Not according to the second small point I just made. They are not systems that become endowed with meaning through *publicly observable use*. They can be observed, but they are not observed in the daily, public life that accounts for the *emergence* of language, the *learning* of language and the *use* of language in communication. Neural nets are crucial for communication and they may correlate with meaning and communication in complicated ways, but they do not have meaning and are therefore not semantic structures.

3. *Holism*: Changes in one part of the structure tend to affect the whole structure, make the parts function differently. Thus, for example, contrasts between elements play an important role in semantics and also in other branches of linguistics. An example of the latter is Roman Jacobson's treatment of phonemes. Many of the structuralists emphasized binary opposition as particularly significant here. However, the emphasis on binary opposition may in part be due to its being easier to spot and easier to handle than more complex structural features.
4. The semantic interpretation is affected by new knowledge: inseparability of theory and theory change on one side and meaning on the other. This is a topic that will be addressed by Alexa McCray and Joon Lee in this volume. This is a main point in Quine: each "translation manual" is a new way of separating theory and meaning, all that matters for correct translation is fit with the publicly observable evidence.

4 Frege on Sense and Reference

Frege contributed more than anybody else to developing semantics.[1] Two of his observations will be particularly important for what follows:

1. *Semantics has two branches:* Theory of sense and theory of reference. This distinction is partly motivated by trying to account for three phenomena: How can identity statements be both true and informative? How can empty names be meaningful? How do we determine what a name refers to?
2. *Frege proposed a unified semantics:* There are three basic kinds of expressions, singular terms, general terms and sentences, which all function in the same way: they have a sense and this sense determines their reference. Carnap's method of intension and extension basically functions in the same way.

5 Quine

We will contrast some of Frege's views with those of Quine and we will see that there are some important points where neither of them can be right.

Quine criticized a notion of *sense* that gave no identity criteria for sense. "No entity without identity" he said in 1957 and he insisted on this point throughout the

---

[1]Particularly important are Frege's *Über Sinn und Bedeutung* [4] and *Der Gedanke* [5]. Wolfgang Künne's recent book, *Die Philosophische Logik Gottlob Freges: Ein Kommentar*, contains some of the main texts and gives a very valuable overview and analysis of Frege's ideas in this field (cf. [9]).

rest of his life.[2] Quine was more satisfied with the theory of *reference*, including a theory of truth. From 1941 on Quine used his views on reference to criticize the modal notions, which he regarded as closely connected with the notion of sense.

Quine presented a number of arguments against the modalities. In the beginning he pointed to their lack of clarity: necessity can be defined in terms of possibility and both notions in terms of possible worlds. However, possible words in turn can hardly be defined without appeal to possibility and thereby we are back in a small circle again, similar to the one that arises in the theory of meaning when we define synonymy and analyticity in terms of meaning and the latter term again in terms of the former. In both cases one moves in a small circle that should satisfy no philosopher.

In later writings Quine reinforced his criticism of the modalities with ever stronger arguments. All of these arguments brought out a severe lack of clarity about what the objects are that we talk about in modal contexts. Quine took this as an indication that there is something fundamentally wrong with the modal notions, like necessity and possibility.

We shall not review these arguments here, but will mention the last of them, which Quine presented in *Word and Object* [11]. Quine here gave an argument that showed that modal distinctions collapse when we quantify into modal contexts. That is, not only is everything that is necessary true, but also conversely, everything that is true, is necessary. So there was no point in the modal notions. Carnap and others had held that without such quantification modal notions would be of no use. So Quine felt that he had clinched his case after 19 years of criticism against the modal notions.

6 Two-Sorted Semantics[3]

Quine's argument was disastrous for the philosophical approach to the modal notions at that time. Formalizing it shows that all assumptions were universally accepted Fregean views on reference, accepted by Carnap and everybody else writing on the modalities at that time. However, the argument is too catastrophic. An analysis of the argument showed that it leads to a similar collapse of all non-extensional notions, not only the logical modalities but also causality, probability, knowledge and belief and legal and ethical notions.

---

[2]Quine used the famous phrase in his presidential address to the Eastern Division of the American Philosophical Association in December 1957, printed in the Association's *Proceedings and Addresses* for 1958, reprinted in *Ontological Relativity and Other Essays* 1969 [12, p. 23]. The idea is clearly there much earlier. Thus in "On what there is" (1948), he writes: "But what sense can be found in talking of entities which cannot meaningfully be said to be identical with themselves and distinct from one another?" [10, p. 4] He used it again in many later writings, for example, in [12, p. 23], [13, p. 102] and [15, pp. 40, 75].

[3]This section and the next are a summary of some sections in my dissertation [3].

So there must be something wrong with the argument. An analysis indicated that the only feasible way of undercutting it was to give up the traditional Fregean view on semantics, the view that singular terms, general terms and sentences all had basically the same kind of semantics. The analysis showed that it can be undercut by a "two-sorted semantics": singular terms behave quite differently from general terms and sentences. This leads us to the so-called "New theory of reference," which we shall now discuss.

7 "New Theory of Reference"

The basic idea in the new theory of reference is the following:

Referring expressions, typically proper names, are "rigid." (This apt expression was introduced by Saul Kripke in his writings on this subject in 1971–1972 [7, 8]. In my 1961 dissertation [3] I called them "genuine singular terms".) They refer to the same object in all different circumstances, or as one would say it in modal logic, in all "possible worlds."

The fact that names and other referring expressions have a radically different semantics from general terms and sentences reflects that objects play a very special and crucial role in our daily lives and in our communication with one another. They are important for our gathering of knowledge and for our actions. This gives them also a special place in semantic theory: One should expect that keeping track of objects is so important that it has given rise to an important feature of language.

We shall now look at the emergence of reference and how it gives rise to this special non-Fregean feature of referring expressions.

8 Three Features of Objects

Objects have *three* features that are crucial for reference:

1. *Transcendence*

First: They are the bearers of a (usually) large number of properties and relations. Normally we know only a small number of these, but the object is conceived of as having numerous *further properties* that we do not know yet, but which are there to be explored. They transcend our knowledge, to use Husserl's phrase. This aspect of objects plays a central role in Husserl's phenomenology, but we shall not go into this here.

2. *Change*

Secondly: Objects, except mathematical ones and a few others, *change* over time. One and the same object can have a property at one time and lack it at another time. The *object* remains identical through changes. Modalities come in at this point; not only are there the actual changes, there are also possible ones, there are accidents and there are necessities. Or, at least, so we say.

3. *Fallibility*

Finally: There is our fallibility. We may have false beliefs about objects. We may seek to correct these beliefs, but all the while our beliefs, true or false, are *of* the objects in question. A belief, or set of beliefs, is not about whichever object happens best to satisfy our beliefs. A semantics that just would seek to maximize our set of true beliefs would reflect poorly the role that objects play in epistemology.

9 Quine on Reification

Starting in *Word and Object* [11] Quine discussed some features of the role of objects in our lives. In the 1990s this became one of his main concerns. Here is an example, from 1995 [16, p. 350][4]:

> As Donald Campbell puts it, reification of bodies is innate in man and the other higher animals. I agree, subject to a qualifying adjective: *perceptual* reification (1983). I reserve "*full* reification" and "*full* reference" for the sophisticated stage where the identity of a body from one time to another can be queried and affirmed or conjectured or denied independently of exact resemblance. [Distinct bodies may look alike, and an identical object may change its aspect.] Such [discriminations and] identifications depend on our elaborate theory of space, time and of unobserved trajectories of bodies between observations.

This distinction between perceptual reification and full reification Quine made already in 1990, although without this terminology. Here is a passage from a lecture Quine gave that year ([14, p. 21], [15, pp. 35–40]):

> I wonder whether a dog ever gets beyond this stage. He recognizes and distinguishes recurrent people, but this is a qualitative matter of scent. Our sophisticated concept of recurrent objects, qualitatively indistinguishable but nevertheless distinct, involves our elaborate schematism of intersecting trajectories in three-dimensional space, out of sight, trajectories traversed with the elapse of time. These concepts of space and time, or the associated linguistic devices, are further requisites on the way to substantial cognition.

10 Three Reasons for Using Names

If I am right in what I have been saying so far, names are normally introduced for the following three purposes:

(i) When we are interested in *further features* of the object beyond those that were mentioned in the description that was used to draw our attention to the object.

---

[4]The italics are Quine's.

(ii) When we want to follow the object through *changes*.

(iii) When we are aware of our *fallibility* and want to refer to an object without being certain that it has some particular property that we can rely on in a definite description.

Given that objects play an important role in our attempts to explore and cope with the world, and given that objects have the features I have listed, we should expect these features to be reflected in our language. We should expect a language to have a category of expressions that is especially designed to refer to these objects and stay with them through all these changes that they and our beliefs about them undergo. This is just what genuine singular terms, or rigid designators, are supposed to do. These terms are hence inseparably tied up with the notions of change and fallibility and not just with the modal notions.

Genuine singular terms comprise the *variables* of quantification and correspondingly the *pronouns* of ordinary language. But also, *proper names* are usually used as genuine singular terms, and sometimes so are even *definite descriptions* in some of their uses. Also *indexicals* and *demonstratives* are genuine singular terms. These latter expressions, like "this" and "now," are used at numerous occasions with ever new references. However, at each occasion of use, the expression is used in order to refer to a specific object, and within that particular situation it is used as other genuine singular terms to keep on referring to that object. Within that situation these terms function like other genuine singular terms, they signal to the listener that we now intend to keep on referring to the same object.

Given our concern with objects and other constancies in the world, we should in fact expect a lot of expressions to have this feature. For example, we should expect this to be the case for mass terms, natural kind terms, properties, etc., as has been pointed out by Kripke and Putnam. Even terms that refer to events will have these features. Events, in spite of their sometimes short duration, are objects that we often want to say several things about and find out more about. However, one must be careful not to believe that all or most terms are rigid. Terms that are not purporting to refer to objects cannot be expected to have this kind of stability. For example is there no reason to expect that general terms will keep their extension from one possible world to the next. On the contrary, if they did, then again modal distinctions would collapse. It is crucial that our semantics be two-sorted; we not only need expressions that keep their reference, but we also need expressions that change their extension.

Fregeans tend to look upon proper names as short for definite descriptions (although in some cases the sense might be embodied in perception and not expressed in language). According to them, names save us from repeating the whole description. They could be called "*names* of laziness," just as Geach talked about "*pronouns* of laziness." There does not seem to be any other role for names for the Fregean. I think that names, like pronouns, are not usually introduced for reasons of laziness. They are introduced for the three reasons I mentioned above, they signal that we are interested in future features of the objects, we want to keep track of it through changes and we know that we may be mistaken about the various features of the object. If we are not moved by any of these three reasons we will normally not introduce names or pronouns.

10.1   Two Examples

Let me illustrate this by two examples. Compare the two descriptions:

the balance of my bank account
the person with the glasses

Here, the first of the descriptions may be by far the most frequently used. On the Fregean view it would therefore be likely to be replaced by a name. However, I doubt that any of you have ever introduced a name instead of this description which you use so often. The reason is that you are not interested in any other feature of this number than that it happens to be the present value of your bank account. If you were particularly interested in other features of that number, perhaps because you are a number theoretician, you would switch to using a numeral.

The second description or others like it, you may, however, have used occasionally in order to pick out an object or a person. If you want to say something more about that person, regardless of whether the person keeps the glasses on or not, then you introduce the person's name, if you know it, or you may use a pronoun, like "he" or "she."

In my view, the notion of a genuine singular term is not fundamentally a modal notion; it is not a notion that requires appeal to necessity or essentialism for its definition or clarification. That genuine singular terms refer to the same object "in all possible worlds" to use modal jargon is not definitory of such terms. It merely follows from the fact that these terms are referring expressions. Preservation of reference is basic to our use of language even outside of modal logic, for example in connection with our talk of an object as changing over time.

11   The Tie Between Word and Object, Causal and Transmission Views

I have argued that we use genuine singular terms to signal our intention to track some particular object through changes, etc. This view is normally taken to mean that genuine singular terms are *guaranteed* to keep their reference. However, the argument for this view in my dissertation is a conditional argument: it says that *if* quantification is to make sense, *then* one has to keep on referring to the same objects "in all possible worlds." There is no guarantee that one succeeds in this, and the argument says that if one gets confused about the references of one's terms, so that they sometimes relate to one object, sometimes to another, then quantification ceases to make sense.

If genuine singular terms were guaranteed to keep their reference, then appeal to a causal tie which connects the term with its object would seem an attractive solution. The tie would hold regardless of confusions and mistakes. Gareth Evans was, I believe, the first to propose such a causal theory of reference [2, pp. 187–208]. The same guaranteed preservation of reference is characteristic of many versions of the transmission theory, which was proposed by Peter Geach and Saul Kripke. Geach is very clear on this. He wrote in 1969 [6, pp. 287–300, pp. 153–165][5]:

> I do indeed think that for the use of a word as a proper name there must in the first instance be someone acquainted with the object named. But language is an institution, a tradition; and the use of a given name for a given object, like other features of language, can be handed down from one generation to another; the acquaintance required for the use of a proper name may be mediate, not immediate. Plato knew Socrates, and Aristotle knew Plato, and Theophrastus knew Aristotle, and so on in apostolic succession down to our own times; that is why we can legitimately use "Socrates" as a name the way we do. *It is not our knowledge of the chain that validates our use, but the existence of such a chain*, just as according to Catholic doctrine a man is a true bishop if there is in fact a chain of consecrations going back to the apostles, not if we know that there is.

However, I regard these accounts as unsatisfactory, for several reasons that I will not discuss here, one of them being their inability to account for reference change.

12   A Normative View on Reference

Let me end by sketching my own view on the tie between genuine singular terms and their objects. Adherents to causal theories of reference and transmission theories of reference of the kind described by Geach tend to regard the tie between singular terms and their objects as a matter of ontology. According to them, we must distinguish between the *ontological* issue of what a name as a matter of fact refers to and the *epistemological* issue of how we find out what it refers to. I look upon the ontological and the epistemological issue as much more closely intertwined. This is largely because language is a *social* institution. What our names refer to—and not only how we find out what they refer to—depends upon evidence that is publicly available in situations where people learn and use language.

Rigidity is not something that is *achieved* through the introduction of a genuine singular term in our language. Sameness of reference is never guaranteed. There is always a risk that in spite of the best of our efforts, we get mixed up, and if the mix-up is persistent and community-wide, the term may even come to *change* its reference. A name does not always continue to refer to the object on which it was bestowed at the original "baptism."

Rigidity, or genuineness, as I see it, is not incompatible with such a reference shift. Instead, I look upon rigidity as an *ideal*, something like a Kantian regulative

---

[5]My italics.

idea, which prescribes the way we use language to speak about the world. When we use a name or another genuine singular term we signal to our listener that we want to refer to a particular object and will do our best to keep on referring to that same object as long as we use the term. There is in our use of names and other genuine singular terms a *normative pull* towards always doing our best to keep track of the reference and keep on referring to it. Sometimes we go wrong and it is unclear both what we believe and what our beliefs are about until a new usage has been established.

All our talk about change, about causation, ethics, and knowledge and belief, as well as about the other modalities, presupposes that we can keep our singular terms referring to the same objects. To the extent that we fail, these notions become incoherent.

To conclude this brief discussion of reference: I hold that there are genuine singular terms, or rigid designators, in our language and that they are indispensable for our talk about change, causality, modality, etc. However, my view differs from other current views on reference mainly in that I do not regard preservation of reference as automatically achieved through our use of singular terms, but as something we *try* to achieve. This is what I mean by "normative pull" and also by what I call a "regulative idea." This view on reference could therefore also be called a normative view on reference.

References

1. Davidson, D.: Truth and meaning. Synthase **17**, 304–323 (1967). Reprinted in Davidson's Truth and Interpretation, pp. 17–36. Clarendon Press, Oxford (1984)
2. Evans, G.: The causal theory of names. Aristotelian Society Supplementary Volume **xlvii**, 187–208 (1973)
3. Føllesdal, D.: Referential Opacity and Modal Logic. Dissertation, Harvard University (1961) (Oslo University Press, 1966, also Routledge, 2004)
4. Frege, G.: Über Sinn und Bedeutung. Zeitschrift für Philosophie und philosophische Kritik **100**, 25–50 (1892). Translated as 'On Sense and Reference' by M. Black in Translations from the Philosophical Writings of Gottlob Frege, 3rd edn., P. Geach and M. Black (eds. and trans.). Blackwell, Oxford (1980)
5. Frege, G.: Der Gedanke. Beiträge zur Philosophie des Deutschen Idealismus I, Heft **2**, 58–77 (1918). Translated as 'The Thought; A Logical Inquiry' by A.M. and Marcelle Quinton in Mind **65**, 289–311 (1956)
6. Geach, P.: The perils of Pauline. Rev. Metaphys. **23**, 287–300 (1969). Reprinted in Geach, Logic Matters, pp. 153–165. Blackwell, Oxford (1972)
7. Kripke, S.: Identity and necessity. In: Munitz, M.K. (ed.) Identity and Individuation. New York University Press, New York (1971)
8. Kripke, S.: Naming and necessity. In: Davidson, D., Harman, G. (eds.) Semantics of Natural Language. Reidel, Boston (1972) (also Harvard University Press, Cambridge, 1980)
9. Künne, W.: Die Philosophische Logik Gottlob Freges: Ein Kommentar. Klostermann Rote Reihe, Frankfurt am Main (2010)
10. Quine, W.V.: On what there is. Rev. Metaphys. 21–38 (1948) [Reprinted in: From a Logical Point of View. Harvard University Press, Cambridge (1953, 1961, 1–19); Harper & Row, Harper Torchbook, New York (1963, pp. 1–19)]

11. Quine, W.V.: Word and Object. MIT, Cambridge (1960)
12. Quine, W.V.: Ontological Relativity and Other Essays. Columbia University Press, New York (1969)
13. Quine, W.V.: Theories and Things. Belknap Press, Cambridge (1981)
14. Quine, W.V.: From Stimulus to Science. Lecture at Lehigh University, Oct 15, 1990 (Manuscript)
15. Quine, W.V.: From Stimulus to Science. Harvard University Press, Cambridge (1995)
16. Quine, W.V.: Reactions. In: Leonardi, P., Santambrogio, M. (eds.) On Quine. Cambridge University Press, Cambridge (1995)

Pragmatism and the Evolution of Semantic Systems

Stefan Artmann

Abstract This paper discusses two classical pragmatists and their theories about the relation of semantics and pragmatics: Peirce's evolutionary pragmatism and its naturalistic reconstruction by Dewey. Two questions are addressed: how do these philosophers understand the origin and development of meaning and how might their theories contribute to an explanation of the evolution of semantic systems? Pragmatism could play a very important role in our understanding of the dynamics of meaning if it integrated theories and models of structural science. This would lead to a new version of pragmatist thought, 'structural pragmatism.'

1 Introduction

Systems, natural or artificial, should be called 'semantic' if their behavior is internally controlled by the processing of signs. The evolutionary explanation of semantic systems of any sort needs a general conception of how meaning (studied by semantics) and behavior (studied by pragmatics) are interdependent. Pragmatism, a philosophical tradition that originated in the USA in the nineteenth century and is currently experiencing a world-wide renaissance, has always been driven on by discussions about the evolutionary dynamics in which objects become meanings through their use in behavioral contexts.

This paper focuses on two classical exponents of pragmatism: Charles S. Peirce and his evolutionary metaphysics and John Dewey and his naturalistic reconstruction of Peirce's ideas. Two questions are addressed: how do these philosophers understand the origin and development of meaning and how might their theories contribute to an explanation of the evolution of semantic systems? These

S. Artmann (✉)
Frege Centre for Structural Sciences, Friedrich Schiller University Jena, Jena, Germany
e-mail: stefan.artmann@uni-jena.de

B.-O. Küppers et al. (eds.), *Evolution of Semantic Systems*,
DOI 10.1007/978-3-642-34997-3_2, © Springer-Verlag Berlin Heidelberg 2013

questions will be approached by discussing Peirce's pragmatic maxim (Sect. 2) and Dewey's theory of the origin of signs (Sect. 3).

The upshot of our discussion is that pragmatist philosophy could play a very important role in the development of an evolutionary understanding of semantic systems if it integrated theories and models of structural science, such as semiotics, cybernetics, and artificial intelligence. The synthesis of pragmatism and structural science would lead to a new version of pragmatist thought, 'structural pragmatism' (Sect. 4).

2 The Pragmatic Maxim in Peirce's Evolutionary Metaphysics

Peirce's pragmatic maxim expresses the core semantic idea of pragmatism *in nuce* by postulating that the meaning of a conception has to be made clear via a systematic analysis of how the object referred to by the conception might interfere with the behavior of the agent who conceives the object. In this section, I first describe how the pragmatic maxim works (Sect. 2.1). Why is the maxim working according to Peirce? To answer this question I introduce his metaphysical concept of evolution and then show what far-reaching assumptions he makes in his evolutionary metaphysics to prove the validity of the pragmatic maxim (Sect. 2.2). Finally, I criticize that, from a pragmatist point of view, Peirce overshoots the mark (Sect. 2.3).

2.1 The Pragmatic Maxim as Semantic Tool

Peirce published the pragmatic maxim in his 1878 paper, *How to Make Our Ideas Clear* [1]. The maxim tells us how we, as rational agents, ought to elucidate the meaning of our conceptions: "Consider what effects, which might conceivably have practical bearings, we conceive the object of our conception to have. Then, our conception of these effects is the whole of our conception of the object" [1, p. 132]. To follow the maxim we must distinguish, in our imagination, a conception from its object. We then imagine what effects that object can cause and focus on such effects which we can imagine that they have practical consequences in the sense that they might influence the actions we could perform. After we developed as precise and complete an idea as possible of those consequences, we know as clearly as possible what our initially vague conception of the object ought to mean: namely, the totality of potential effects of the object on our possible actions.

For Peirce, the pragmatic maxim is a general tool for the semantic analysis of conceptions. He suggests using the maxim, not only to make clear ideas that obviously are practically relevant, but also to elucidate the most abstract philosophical conceptions. In *How to Make Our Ideas Clear*, Peirce discusses, as an example, the idea of reality (see [1, p. 136 ff.]). This should not come as a surprise: by distinguishing a conception from its object, the pragmatic maxim implicitly presupposes an idea of reality, which must be made clear if we want to fully understand our conception of the pragmatic maxim (such an idea may be called 'pragmatist category'). Peirce initially suggests defining the real to be "[. . .] that whose characters are independent of what anybody may think them to be" [1, p. 137]. Thus, the maxim's distinction of a conception from its object turns out to be the distinction of what is dependent on imagination from what is not dependent on it. But how can we conceive of that difference?

Even in case of pragmatist categories, we cannot but follow the pragmatic maxim in trying to understand what they mean as clearly as possible. The circularity of that undertaking is, for a pragmatist, a virtuous one: it shows that we, as rational agents, cannot transcend the complex interdependence of thought and action.

What does result from applying the pragmatic maxim to the idea of reality that is implied by the very same maxim? If the clearest conception we can have of an object and its properties is the conception of the totality of its potential effects on our possible actions, then the idea that an object is real must be a subset of that totality. According to Peirce's provisional definition cited above, to that subset belong all effects whose properties are independent of any conception of them. Yet how can we come to know which effects are real in that sense? Peirce's answer is (see [1, p. 138 f.]) that all rational agents have to start a joint research through which their conceptions will be experimentally tested. By following the pragmatic maxim, those conceptions are tested for their ability to foresee the practical effects of the objects they refer to, on the actions the rational agents perform. In this way, the pragmatic maxim acts as a semantic tool: it induces us to test the reality of the objects of our conceptions by using the latter intentionally as guide-posts in planning and realizing our interactions with the objects of our conceptions.

According to Peirce, the experimental investigation of causal relations by which the objects of our conceptions interact with our actions would clean off any idiosyncrasy from the conceptions we would more and more extensively share with each other. Pushing this process to its logical end, we are allowed to hope that it converges necessarily on absolutely clarified, or true, conceptions whose objects are acknowledged as being real by all researchers: "The opinion which is fated to be ultimately agreed to by all who investigate, is what we mean by the truth, and the object represented in this opinion is the real" [1, p. 139]. The real is, for Peirce, neither the totality of what is given here and now, nor the thing-in-itself which we can never know. The real is what shows itself in our conceptions at the end of our common experimentation. The practically most relevant effect of the object of Peirce's conception of the real is to let our actions be guided by the hope that our conceptions, by experimenting with them, will be purified from idiosyncrasies of mine and yours.

2.2   Evolutionary Theory as Metaphysical Support of the Pragmatic Maxim

Peirce argues for his conception of reality and his hope that we shall know the real, by trying to give us an absolutely reliable guarantee that we are not hoping in vain. Peirce's reasoning takes on the form of an evolutionary metaphysics. Why? Or, as a pragmatist would ask, what is the practical relevance of the object of Peirce's conception of evolution? Let us start, for example, by vaguely imagining a process in which neither is the same repeated again and again nor do ever new events succeed each other in a completely random fashion.[1] An evolutionary process connects regularity and randomness so that we can observe continuous transformations of existent process components. Taking part in such an evolution, we may suppose that we perform our actions in an environment that is stable over certain stretches of time so that, e.g., even catastrophic events do not lead to a sudden change in fundamental causal laws. We may suppose, too, that our actions can intentionally change our environment with rather predictable results and that we usually have time enough to adapt to unforeseeable environmental changes. In short, we can develop habits in such an ambience and can change it to adjust it to our habits.[2]

If the cognitive and behavioral evolution of the community of rational agents is part of the evolution of what is real, then those rational agents are justified to conceive, first, that they can learn continually from their unsuccessful actions and adapt themselves to reality, and second, that they can adapt, in a certain degree, what is real to their conceptions. For Peirce, the conditional clause of the preceding sentence must express a truth. Otherwise, his pragmatic maxim would lead to philosophical confusion: it could then not be applied to its components (here, the conception of reality) in order to make our conception of the whole maxim clear.

Peirce's attempt to prove the reality of learning and adaptation can be reconstructed succinctly as an evolutionary narrative in seven chapters.

1. *From random to iconic events.* To get started ex nihilo, an evolutionary process depends on events that are random in an absolute sense: they are completely unforeseeable and thus uncontrollable (cf. Peirce's speculative concept of the First and its cosmogonic interpretation as the "womb of indeterminacy" in [3, p. 273 ff.]). Such events can be conceived of only after they have occurred. However, if similar events are happening, any of those events is representing all events which are related to each other by their similarity. Those events are not absolutely random anymore since some of their features are known before they happen. Using Peirce's semiotic terms, we can say that such events are 'iconic'

---

[1]Peirce criticizes evolutionary theories that, as he sees them, overemphasize either the role of necessity (e.g., August Weismann's) or the role of chance (e.g., Charles Darwin's), in his 1893 paper, *Evolutionary Love* (see [2, p. 358 ff.]).

[2]For Peirce, the main evolutionary principle is thus neither necessity nor chance but habit (see [2, p. 360 ff.]).

ones: "I call a sign which stands for something merely because it resembles it, an *icon*" [4, p. 226].

2. *From iconic to indexical events.* It might occur that, when an iconic event of similarity class i_1 is happening, it is sometimes the case that an iconic event of similarity class i_2 is happening shortly thereafter. Both events are then spatio-temporally connected to each other in a particular way. Using Peirce's semiotic terminology, such events are indexical representations of each other (see [4, p. 226] for the definition of indexical sign, which "[...] signifies its object solely by virtue of being really connected with it."). The more often an i_1-event is happening under circumstances that are also similar to each other, the more precisely the probability that an i_2-event is happening shortly thereafter will become fixed. Thus the following generalization becomes possible: "[...] *experiences whose conditions are the same will have the same general characters*" [5, p. 169].

3. *From indexical to symbolic events.* Why does the probability that an i_2-event is happening given an i_1-event under particular circumstances of similarity class c get fixed more and more precisely? After very small regularities between indexical events appeared, there emerges a new type of events that are rule-based or, in Peirce's semiotic terms, 'symbolic' (see [4, p. 225 f.] on tokens or symbols). The rules are of the following form: If it is more probable that an i_2-event is happening after an i_1-event happened under c-circumstances than when an event of any other similarity class did, and if an i_2-event is desired to happen, then the given situation shall be transformed to c-circumstances in a way that makes an i_1-event most probable. Knowledge of such rules is an epistemic precondition of becoming a rational agent who follows those rules to effectuate events in a controlled fashion. Rational agents conceive of their actions as symbolic representations of indexical relations of events and they develop the habit of performing particular actions, given specific situations, in order to bring about particular indexical relations (see [3, p. 257 ff. and p. 262 ff.] on synthetic consciousness and habit, respectively). In this way, rational agents proactively adapt the environment in which they act, to their desires so that they are controlling it better and better. By so doing, they reorganize their ambience so that the probability that an i_2-event is happening given an i_1-event under c-circumstances gets fixed more and more precisely.

4. *Evolution as symbolization.* Peirce calls the process in which events are becoming symbolic "[...] that process of evolution whereby the existence comes more and more to embody those generals which were [...] said to be *destined*, which is what we strive to express in calling them *reasonable*" [6, p. 343]. The definition of evolution given above as a process in which regularity and randomness are so related that we may expect to be acting in a stable world, can now be extended. Symbolization means, for Peirce, the establishment of regularities by rule-following actions. When more and more events are becoming symbolic, the general relation of the regular and the random is an evolving one, too. The rules that are effective in an evolutionary process themselves are resulting from evolution. According to Peirce, this is valid not only for the rules we are following in our actions. The physical laws of nature, too, have been developing

in cosmic evolution and can be changing further (see [3, p. 279]). Since Peirce does not allow for any regularity that is not explainable evolutionarily, he must understand the beginning of an evolutionary process as a 'state' of unbounded irregularity—more precisely, as an absolutely random event (see [3, p. 278]).

5. *Evolution as meta-symbolization.* Yet why did the absolutely random origin of all that really exists, start an evolutionary trend towards symbolization, i.e., ever growing regulation? This trend must be seen as the realization of a rule that is, in a sense, applying itself to itself necessarily, and is amplifying its own validity as soon as it randomly starts to become effective at some time:

> [...] in the beginning,—infinitely remote,—there was a chaos of unpersonalised feeling, which being without connection or regularity would properly be without existence. This feeling, sporting here and there in pure arbitrariness, would have started the germ of a generalizing tendency. Its other sportings would be evanescent, but this would have a growing virtue. Thus, the tendency to habit would be started; and from this with the other principles of evolution all the regularities of the universe would be evolved. At any time, however, an element of pure chance survives and will remain until the world becomes an absolutely perfect, rational, and symmetrical system, in which mind is at last crystallized in the infinitely distant future. [7, p. 297]

Thus, the self-applying meta-rule *Become symbolic!* asserts itself over absolute randomness. The evolutionary trend towards symbolization will become symbolized itself sometime. Of course, Peirce sees himself as the philosopher who achieves this feat: he considers his philosophy the generally valid conception of evolving reality, to which scientific research will contribute the details (see [3, p. 247] and [7, p. 297]).

6. *God as foundation of evolution.* We can apply the pragmatic maxim onto itself in order to test its practical consequences—so we started our exploration of Peirce's metaphysics. But in case of the fundamental principle of evolution doubts might arise: how can it be self-applying and ensure its own effectiveness? Peirce's answer specifies a being that we are justified to ascribe such a self-founding power to: God (see, e.g., concerning God as foundation of our evolving knowledge of reality: "[...] God's omniscience, humanly conceived, consists in the fact that knowledge in its development leaves no question unanswered" [8, p. 236]). The fundamental principle of evolution is, in a sense, the pragmatic maxim writ in large that the creator of the evolving universe is following: God's conceptions are evolutionarily realized in creation by rational agents who are giving God's conceptions real meanings and are making God's conceptions clear by partaking in creation.

7. *In God we trust ourselves as rational beings.* Wherefrom do we, as rational agents, know of the existence of God in whose creation we are cocreatively acting? Peirce appeals to us that we bring ourselves into a state of 'musement,' i.e., of non-intentional, meditative observation and thought (see [9, p. 436 ff.]). Peirce is prophesying that in such a state we conceive of the necessarily vague hypothesis of the reality of a creative infinite being we cannot fully understand: God. Due to that hypothesis we develop an ardent love to reality as the totality of what has been created in a way that allows us to realize our rational conceptions

by acting cocreatively according to them. Symmetrically we experience the love of the creative God towards us, his creatures, as a metaphysical encouragement to cocreatively realize our rational intentions. The object of our conception of God has, thus, the fundamental practical effect that we will, more and more intensely, strive to act rationally, particularly by using the pragmatic maxim to make our conceptions as clear as possible. Since we, as rational agents, must greet that effect of our belief of God (otherwise, we would contradict our own rationality), we can assert rationally that God exists (see [9]).

So the evolutionary story of Peirce's pragmatist proof of God's existence, of the possibility that we can know what is real, and of the validity of the pragmatic maxim, goes.

2.3 A Pragmatist Criticism of Peirce's Evolutionary Metaphysics

Peirce's evolutionary speculation about the reason why the pragmatic maxim is valid ends in an appeal to experience the universe meditatively as God's creation. Now we may wonder and ask: does Peirce direct his religious appeal to the same addressee as he does recommend his pragmatic maxim to?

Peirce's metaphysics of evolution leads to an absolute version of the pragmatic maxim: the evolution of what is real and the possibility of its being known are based upon a creator who explicates, in his creation, the practical consequences of the objects of his conceptions. Those consequences are nothing but evolving reality and become known as real by rational agents. In this sense, Peirce's idea that the community of rational agents progresses towards true conceptions, so that the real will have been symbolized at the end of their research, is essentially an eschatological one: rational agents will finally know what has been, as vague conceptions, in the mind of God at creation by conceiving, as cocreators, what is real as clearly as possible and controlling their actions accordingly.

The original addressee of Peirce's maxim is a rational agent who wants to have clear conceptions. Peirce does not search for the foundation of his maxim by applying it to itself in the scope of its procedure of making conceptions clear, i.e., in respect to possible actions of rational agents. Instead, Peirce's self-application of the pragmatic maxim proceeds by speculations that elevate conceptions needed for the explication of the maxim into an absolute form. Peirce is not focused anymore on the relation of our everyday and philosophical conceptions to the world in which we act. He is not focused anymore on the not completely controllable environment by which we are brought to think about the potential effects of the objects of our conceptions on our possible actions. And he is not focused anymore on the causal relations, which we explore experimentally in stretches of time we can control to some degree, between the objects of our conceptions and our actions.

Peirce lets himself be swept away by his absolutist leanings and addresses his evolutionary metaphysics to an observer who is musing about the cosmos as a whole. Such a meditative, Cartesian program of philosophy Peirce does criticize vehemently—as he is obliged to do as a pragmatist—in earlier texts as a project of philosophical research that is doomed to failure from the beginning (see, e.g., [1, p. 125 ff.], on belief in a Cartesian and in a pragmatistic sense). But when he is realizing his own pragmatist program, Peirce is adapting, by speculating about the metaphysical foundation of his pragmatic maxim, to a style of philosophizing that would be quite satisfying for a philosopher in the Cartesian tradition. To its original addressee the pragmatic maxim hands over a semantic tool that helps her to make her conceptions clear and her actions rational. For that, it is not necessary that the maxim gives her a guarantee that conceptions made clear, and actions made rational, by the maxim are clear and rational *sub specie aeternitatis* because they are contributing to making real, and known, the creative intentions of God.

3 The Reconstruction of Meaning in Dewey's Evolutionary Naturalism

The decisive strategic difference between Peirce's and Dewey's pragmatism is in the determination by which they are anchoring their philosophies in the problematic situation of rational agents who cannot control their environments completely. Whereas Peirce deliberately loses track of the original addressee of his pragmatic maxim, Dewey struggles to keep even his most abstract argumentations tightly connected to situations in which agents try to solve problems of life rationally by their finite means. That strategic difference accounts also for Peirce's and Dewey's diverse use of evolutionary thinking in the explanation of meaning. To show this, I first describe Dewey's method of pragmatist reconstruction as a transformation of Peirce's pragmatic maxim (Sect. 3.1). Then I exemplify how Dewey uses evolutionary theory as naturalistic support of pragmatism by analyzing his concept of language as "the tool of tools" [10, p. 134] (Sect. 3.2). Finally, I sketch a pragmatist criticism of a dysfunctional Peircean residue in Dewey's pragmatism (Sect. 3.3).

3.1 Pragmatist Reconstruction as Semantic Instrument

According to Peirce's pragmatic maxim, a conception that contributes to our knowledge of reality must have an object that can affect, in some way, our possible actions. Such an object is represented in its effects that interfere with our actions; its effects are indices of the object. Our conception is then a symbolic representation of the indexical relation between the object and its effects on our actions. If, for example, an action did not reach its intended aim, we change—then and now, at least—our conceptions of the objects implicated in the action because we want to understand what went wrong. Peirce's pragmatic maxim is, thus, a postulate that we should make our symbolic representations of the world clear by exploring the indexical relations of its objects to their effects on our actions.

Dewey defines his own version of pragmatism, which he calls 'instrumentalism', in a way that shows, on the one hand, how close his thought is to Peirce's, and indicates, on the other, where he deviates from Peirce:

> Instrumentalism is an attempt to establish a precise logical theory of concepts, of judgments and inferences in their various forms, by considering primarily how thought functions in the experimental determinations of future consequences. That is to say, it attempts to establish universally recognized distinctions and rules of logic by deriving them from the reconstructive or mediative function ascribed to reason. [11, p. 9]

Peirce develops a general theory of meaning, knowledge, and reason from an evolutionary perspective on their role in creative action. Dewey, too, asks how the forms of thinking are determined by their anticipatory functions in processes of changing the world. But, by calling his pragmatism 'instrumentalism', Dewey suggests that his philosophy is based on a conceptual structure in which the pragmatist categories of means and ends are central. And, by calling the principal function of thought "reconstruction," he suggests that any explanation of what meaning, knowledge, and reason are must take their transformational nature very seriously.

Analogously to Peirce's category of reality as conceptual presupposition of the pragmatic maxim, Dewey's categories of means and ends are logical tools that are needed for making the functioning of reason as clear as possible. According to Dewey, the world is not ontologically divided into means and ends. Anything might function as a means in some contexts where rational beings are acting, and as an end in other such contexts: "The distinction of means and end arises in surveying the *course* of a proposed *line* of action, a connected series in time. The 'end' is the last act thought of; the means are the acts to be performed prior to it in time" [12, p. 27]. This is also true when reason is trying to understand its own functioning, e.g., when applying the pragmatic maxim to the conceptions it generates.

For Dewey, thought and actions both are means as well as ends in reconstruction. 'Reconstruction' does not mean imitation of a lost original but reorganization of an existent object so that the result fulfils a particular function in a satisfactory way. This function might have been fulfilled worse by the initial object. Yet the function of the resultant object could also have been changing during the process of reconstruction. It is a special case when the function of the resultant object is to create the impression to be identical to another, maybe lost entity.

Dewey considers anything that is involved in thinking and acting, as being accessible to the same method of reconstruction. He is a monist in all things methodological. Reconstruction, the use of thought for 'the experimental determination of future consequences,' is as relevant to the understanding of reason and culture as it has already been for the exploration of nature since Galileo and Newton (see [13, p. 18]). Pragmatist reconstruction, "the method of observation, theory as hypothesis, and experimental test" [13, p. 10], is the semantic instrument by which Dewey grapples with any problem of meaning, knowledge, and reason for the purpose of letting reason understand itself as *the* means of experimentation.

3.2 Evolutionary Theory as Naturalistic Support of Pragmatist Reconstruction

Peirce's evolutionary metaphysics has the function to prove that the key insight of pragmatism, which is expressed in the pragmatic maxim, about the intrinsic relation of meaning and action is true. His speculations about the evolving cosmos, however, lose sight of the primary addressee of pragmatism: the rational agent who tries to make her conceptions clear in a world she cannot control completely by her own finite means. Dewey, in contrast, recommends evolutionary thinking, more precisely: Darwinian thinking, to the rational agent, who followed philosophers such as Peirce, so that she can rediscover the specific circumstances under which she tries to make her conceptions clear. For Dewey, Darwin's main scientific as well as philosophical merit is to have transferred the experimental method of modern science to research on organisms, including human behavior and consciousness. Darwin achieved this feat by destructing "the dogma of fixed unchangeable types and species, of arrangement in classes of higher and lower, of subordination of the transitory individual to the universal or kind" [13, p. 76]. The Darwinian theory of the origin of species by natural selection explains how general forms, which are called 'species,' evolve through the competitive and cooperative interactions of individual organisms. Without falling back into simply registering a myriad of single facts, Darwin introduced, into biology, a genetic logic of the context-dependent formation of the general from the individual (see [14, p. 41]).

Stated as abstractly as possible, Darwin's insight is that "[t]here is [...] such a thing as relative–that is *relational*–universality" [13, p. 13]. Philosophy, Dewey challenges his fellow pragmatists, has to develop as comprehensive an understanding of that insight as is possible for human reason. Philosophy would thus become the science of mechanisms through which generalizations are developed under particular circumstances and would use those mechanisms in order to explain why people interpret their life as they do and to give them the means through which they can meliorate their situation according to their own preferences (see [14, p. 43 f.]).

From Peirce: Dewey adopts the general name for generative mechanisms when phenomena of meaning, knowledge, and reason are concerned, he calls them 'habits.' As to humans, Dewey defines 'habit' as "[...] activity which is influenced by prior activity and in that sense acquired; which contains within itself a certain ordering or systematization of minor elements of action; which is projective, dynamic in quality, ready for overt manifestation; and which is operative in some subdued subordinate form even when not obviously dominating activity" [12, p. 31]. Habits are mechanisms that produce behavioral patterns which transcend a particular context but are context-sensitive in the sense that they have been developed in the life history of a human being and are thus actualized only if a context is experienced, by that individual, as meeting certain conditions.

For Dewey, evolution of life consists in the transformation of habits leading to more and more flexible and powerful behavioral patterns and to environments that are controlled better and better (see [13, p. 82]). This includes also human reason, which is not only a product of evolution but also a powerful motor of evolution, since it is the most advanced means by which existent circumstances can be transformed and controlled (see [15, p. 123]). All aspects of human reason—including its most potent habit, the method of modern science (see [16, p. 99])—are evolutionarily explainable phenomena. It does not come as a surprise that Dewey wants also to give an evolutionary account of the invention of language, which he considers the most important event in the history of humanity (see [16, p. 121]). Language is of utmost relevance also to an ontogenetic understanding of individual rational agents since linguistic structures constitute the fundamental categories of human cognition (see [13, p. 86 f.]). Reason, or to use the designation preferred by Dewey: intelligence, is nothing but the habit to regard an object as representing another one, i.e., as a sign or symbol (see [16, p. 170]).

In one of his chief works, *Experience and Nature*, Dewey tells a story of how symbols originate through the interaction of humans. Neither do signs make up the material traces of spiritual entities, such as Platonian ideas, nor do they result from the expression of mental states, such as pain or joy. Things become symbols when they are marked, by interacting humans, as having particular functions in the performance of social behavior (see [10, p. 138]). To symbolize something objects first must be made meaningful by human gestures and sounds that have been becoming more than just physical events (see [10, p. 138 f. and p. 142]). How does this come about? Here is Dewey's story:

> *A* requests *B* to bring him something, to which *A* points, say a flower. There is an original mechanism by which *B* may react to *A*'s movement in pointing. But natively such a reaction is to the movement, not to the *pointing*, not to the object pointed out. But *B* learns that the movement *is* a pointing; he responds to it not in itself, but as an index of something else. His response is transferred from *A*'s direct movement to the *object* to which *A* points. Thus he does not merely execute the natural acts of looking or grasping which the movement might instigate on its own account. The motion of *A* attracts his gaze to the thing pointed to; then, instead of just transferring his response from *A*'s movement to the native reaction he might make to the thing as stimulus, he responds in a way which is a function of *A*'s *relationship*, actual and potential, to the thing. The characteristic thing about *B*'s understanding of *A*'s movement and sounds is that he responds to the thing from the standpoint of *A*. He perceives the thing as it may function in *A*'s experience, instead of just ego-centrically. Similarly, *A* in making the request conceives the thing not only in its direct relationship to himself, but as a thing capable of being grasped and handled by *B*. He sees the thing as it may function in *B*'s experience. Such is the essence and import of communication, signs and meaning. Something is literally made common in at least two different centres of behavior.

> To understand is to anticipate together, it is to make a cross-reference which, when acted
> upon, brings about a partaking in a common, inclusive, undertaking. [10, p. 140 f.]

In the terminology of Peirce's semiotics, we can say that, for Dewey, the origin of signs is in the symbolization, by a human being (B), of indexical relations (pointing) between human actions (advancement of A's arm towards object) and physical objects (flower). 'Symbolization' means that B learns to consider the object to which A has an indexical relation as being in that relation. Then the object is seen, by B, as an object that is a means for A to involve B into A's action so that one of A's ends may come into being with B's help. Thus, the object, as a communicative tool, has been becoming meaningful to A and B (see [10, p. 146 f.]). Being symbolized, the indexical relation of A to the object has become more than just a physical incident: first, A and B have developed a common understanding of the object as tool and, second, both A and B can symbolically use the same form of indexical relation in other contexts in order to communicate the same type of meaning to each other. Meanings are generalizing methods of action, or habits, which philosophers tend, as so-called 'ideas,' to cut off from "their naturalistic origin in communication or communal interaction" [10, p. 149].

To characterize symbolization in terms of means and ends, Dewey uses the concepts of the instrumental and the consummatory. An object used as a symbol stands generically for something that does not exist yet: it is thus instrumental to future consequences of its being used as a symbol. However, the very same object still is a physical entity that can become, e.g., the object of esthetic perception by which its sensual properties are consummatorily experienced in themselves (see [16, p. 188]). The same duality of the consummatory and the instrumental applies to social action and thus to communication, which is both a means to an end and an end in itself: "For there is no mode of action as fulfilling and as rewarding as is concerted consensus of action. It brings with it the sense of sharing and merging in a whole" [10, p. 145]. Communication, by giving everyone in a community the chance to participate in the shared understanding and common development of habits that are socially valued, is even the highest end in itself (see [10, p. 157 and p. 159]). Dewey's emphatic characterization of the consummatory character of communication as a deeply meaningful "communion" [10, p. 159] lets him write that "[i]n communication, such conjunction and contact as is characteristic of animals become endearments capable of infinite idealization; they become symbols of the very culmination of nature" [10, p. 157]. From here it is just one small step to a religiously sounding praise of communication: "[. . .] when the emotional force, the mystic force one might say, of communication, of the miracle of shared life and shared experience is spontaneously felt, the hardness and crudeness of contemporary life will be bathed in the light that never was on land or sea" [13, p. 164]. Such an emphatic promise of salvation indicates a principal problem that a pragmatist ought to have with Dewey's evolutionary theory as naturalistic support of instrumentalism.

3.3 A Pragmatist Criticism of Dewey's Evolutionary Naturalism

Dewey's evolutionary theory of reason culminates in the proposition that communication is the action through which human beings symbolize their rational capabilities by means of social interaction. Coming together in order to talk with each other becomes a symbol of joining a reasonable community that strives for the joint realization of shared aims. Yet communication itself, as the most deeply human form of rational action, is as natural a process as any other intelligent activity: "The intelligent activity of man is [. . .] nature realizing its own potentialities in behalf of a fuller and richer issue of events" [16, p. 171]. Humans, by acting as rational beings, use nature as a means in order to realize nature as an end. Nature is a means if it is considered as the sum total of existent conditions that can be used to make human aims real. Nature is an end if it is considered as the sum total of possible conditions that would allow human beings an ever more rational way of living. Nature as an end instructs the reconstruction of nature as a means whenever the latter is "perfected" [16, p. 241] by humans, i.e., whenever the existent state of nature is transformed in such a way that it is becoming a better support of realizing rational aims.

What is the standard against which the rationality of human aims is measured so that, in turn, existent nature as a means can be judged whether it ought to be transformed? For Dewey, such standards are originating in nature as an end: they express, in form of values, the conditions under which natural entities could fulfill their functions in the best way possible (see [16, p. 207 f. and p. 212]). The highest standard of rationality must be anchored in the end of nature as a whole—which is, Dewey proposes, nature itself as an infinite process of perfecting itself: "The end is no longer a terminus or limit to be reached. It is the active process of transforming the existent situation. Not perfection as a final goal, but the ever-enduring process of perfecting, maturing, refining is the aim in living. [. . .] Growth itself is the only moral 'end' " [13, p. 141].

Yet how can the original addressee of pragmatist reconstruction, the finitely rational agent, know which of her possible actions is contributing the most to the process of perfecting nature, which is an end in itself? Dewey offers one principal answer: the more an action contributes to the overcoming of divisions that disturb communication, the more rational it is. Why? First, the all-encompassing imagination of the evolutionary unity of nature is the unsurpassable symbol for any process in which reason emerges and through which its aims can be realized (see [17, p. 85]). Second, Dewey interprets that holistic symbol with respect to the evolutionary connectedness of rational agents and elevates unification through communication to the status of the fundamental principle of reasonable action.

Dewey's symbolization of evolving nature and progressive growth, his naturalistic holism, fulfils the same function as Peirce's speculative proof of the existence of God. Imagining the unity of nature as symbol for the indexical relation of each rational being to the totality of conditions and results of his thoughts and actions is, for Dewey, the best way to make us believe that our rational habits will lead to an improvement of our lives. Being guided by a firm belief in the unity of nature—which belief is the functional equivalent of Peirce's musement of God—, the individual human considers herself an instrument of perfecting that unity.

> [...] I question whether the spiritual life does not get its surest and most ample guarantee when it is learned that the laws and conditions of righteousness are implicated in the working processes of the universe; when it is found that man in his conscious struggles, in his doubts, temptations, and defeats, in his aspirations and successes, is moved on and buoyed up by the forces which have developed nature; and that in this moral struggle he acts not as a mere individual but as an organ in maintaining and carrying forward the universal process. [18, p. 235]

Dewey defines his pragmatist category of God as denoting "the unity of all ideal ends arousing us to desire and actions" [17, p. 42]. Made clear by applying the pragmatic maxim, his concept of God refers to the feeling that a rational agent has when letting her actions be guided by her imagination of the unity of nature. "It is this active relation between ideal and actual to which I would give the name 'God'" [17, p. 51]. For Dewey, we literally are evolving God if we reconstruct nature in order to perfect it. We then feel the community of rational agents to be the driving force of evolution. Dewey's naturalistic holism thus becomes indistinguishable from what may be called 'community pantheism.'

Dewey applies his evolutionary naturalism to nature as a whole and weakens his strategy to keep even his most abstract argumentations tightly connected to situations in which agents try to solve problems of life rationally by their finite means. In analogy to my pragmatist criticism of Peirce's evolutionary metaphysics, I propose that a pragmatist should give a negative answer to the question whether Dewey directs his pantheistic appeal to the same addressee as he does recommend his method of pragmatist reconstruction to. Dewey's original addressee, a finitely rational agent, wants to understand how she can reconstruct the environment she lives in, according to preferences that should meet criteria of rationality. Yet the addressee of Dewey's holistic version of naturalism is an agent who yearns for an unshakable feeling that unfailingly motivates her to act rationally. To such an agent, Dewey supplies his advice to interpret social communication as a secular ersatz of Holy Communion. If she accepts Dewey's offer, the agent regards herself as a tool of evolving nature. From rational reconstruction, via social fusion, to cosmic religion: this must not be the way of a philosophical program that shall help individual rational agents to make their conceptions clear and their actions rational, by developing a pragmatist understanding of meaning, knowledge, and reason.

4 Structural Pragmatism and the Evolution of Semantic Systems

The genuine starting point of any pragmatist inquiry into meaning is the situation where a rational agent is trying to solve problems, by her finite means, in an environment she cannot control completely. The reason why Peirce and Dewey went wrong on their way towards an evolutionary theory of meaning is that both philosophers tried to find some absolute point of reference to be reached from their starting point. For doing so, they paradoxically used evolutionary reasoning, which, we would rather suspect, proceeds towards a relativistic stance on all questions philosophical. Peirce wanted to prove that our conceptions will lead, in the long run, to our knowing the real since we are cocreating the rational world by experimenting with our conceptions. The evolutionary process as realization of truth is, as a whole, Peirce's absolute point of reference. Dewey showed a moderated, yet analogous, tendency: he wanted to provide his followers with an irresistible motivation to act rationally and did so by appealing to the socio-cosmic emotion of being an instrumentality of evolutionary nature as a whole. Yet the finitely rational agent, whom the pragmatist philosopher tries to help solve her theoretical and practical problems, does originally ask neither for an unshakable foundation of her rational conceptions nor for an overwhelming motive for acting rationally.

However, Peirce's pragmatic maxim expresses the basic method of any pragmatist inquiry into semantics. And Dewey's pragmatic reconstruction is the principal strategy for any pragmatist who sees continuity between scientific and philosophical approaches to rationality. Even beyond methodology, Peirce's and Dewey's evolutionary thinking is full of important contributions to an understanding of semantic systems. Peirce's semiotics, for example, provides a systematic framework for the description of sign processing. And Dewey's insight that an evolutionary understanding of meaning is in need of a general logic of context-dependence still is of highest pertinence.

Yet how can we make as best a use of those pragmatist contributions to the evolutionary theory of semantic systems as possible? By interpreting them as abstract, yet not formalized, descriptions of basic structures that determine any relation, theoretical or practical, of an agent to its environment. By 'structure,' we mean general relational forms that are not restricted to be found in a particular empirical field of research, but may occur in many diverse fields. A scientific research program that analyzes a wide variety of empirical phenomena by means of a theory of some structure is called 'structural science' (see [19–21]). For example, Peirce's structural concept of sign as general relation of representation may apply not only to human languages but also to animal communication systems.

From a Deweyan perspective, each structural science addresses a particular aspect of habit, which is the most general and comprehensive relation of agents and environments, and develops formal models of that aspect with the structure of habit serving as a vanishing point. When Dewey sketches, for example, the sensorimotor coordination of an organism in its environment, we are reminded of the description of a cybernetic feedback loop *avant la lettre* (see Dewey's 1896 paper, *The Reflex Arc Concept in Psychology* [22]). His remarks on context-dependence should be connected with Bernd-Olaf Küppers' project of developing a structural science of boundary conditions, called 'peranetics' (see [23, p. 314]). Dewey's general picture of human rationality and Herbert A. Simon's structural science of bounded rationality deeply agree on the necessity of searching for "[...] relative invariants: regularities that hold over considerable stretches of time and ranges of systems"

[24, p. 36]. More examples of research programs of structural science that address particular aspects of Dewey's general concept of habit can easily be found.

Analyzing the situation where a rational agent is trying to solve problems, by her finite means, in an environment she cannot control completely, a pragmatist philosophy may heavily be using the methods, theories, and results of structural sciences. If, by doing so, that philosophy tries also to make the concepts of structural sciences clear, then I propose to call such a philosophy 'structural pragmatism.' One of the most important differences of structural pragmatism and Dewey's instrumentalism is that the former, while interpreting structural science in the light of the relation of agents to their environment, does not interpret that relation, or any of its aspects, in an absolutist manner. Though the interaction of agents and their environments is structured by what Dewey calls 'habits,' the cosmos, as the totality of environments past, present, and future, should not be seen as a total habit evolving, if we want to remain pragmatist philosophers and do not want to turn into theologians who go, a little bit too easily, over the boundary between the actions of finite beings and the 'doings' of an infinite being.

References

1. Peirce, Ch.S.: How to make our ideas clear. In: Houser, N., Kloesel, N. (eds.) The Essential Peirce. Selected Philosophical Writings, vol. 1 (1867–1893), pp. 124–141. Indiana University Press, Bloomington (1992)
2. Peirce, Ch.S.: Evolutionary love. In: Houser, N., Kloesel, N. (eds.) The Essential Peirce. Selected Philosophical Writings, vol. 1 (1867–1893), pp. 352–371. Indiana University Press, Bloomington (1992)
3. Peirce, Ch.S.: A guess at the riddle. In: Houser, N., Kloesel, N. (eds.) The Essential Peirce. Selected Philosophical Writings, vol. 1 (1867–1893), pp. 245–279. Indiana University Press, Bloomington (1992)
4. Peirce, Ch.S.: On the algebra of logic: a contribution to the philosophy of notation [Excerpts]. In: Houser, N., Kloesel, N. (eds.) The Essential Peirce. Selected Philosophical Writings, vol. 1 (1867–1893), pp. 225–228. Indiana University Press, Bloomington (1992)
5. Peirce, Ch.S.: The probability of induction. In: Houser, N., Kloesel, N. (eds.) The Essential Peirce. Selected Philosophical Writings, vol. 1 (1867–1893), pp. 155–169. Indiana University Press, Bloomington (1992)
6. Peirce, Ch.S.: What pragmatism is. In: Peirce Edition Project (ed.) The Essential Peirce. Selected Philosophical Writings, vol. 2 (1893–1913), pp. 331–345. Indiana University Press, Bloomington (1998)
7. Peirce, Ch.S.: The architecture of theories. In: Houser, N., Kloesel, N. (eds.) The Essential Peirce. Selected Philosophical Writings, vol. 1 (1867–1893), pp. 285–297. Indiana University Press, Bloomington (1992)
8. Peirce, Ch.S.: An American Plato: review of Royce's 'Religious aspect of philosophy'. In: Houser, N., Kloesel, N. (eds.) The Essential Peirce. Selected Philosophical Writings, vol. 1 (1867–1893), pp. 229–241. Indiana University Press, Bloomington (1992)
9. Peirce, Ch.S.: A neglected argument for the reality of God. In: Peirce Edition Project (ed.) The Essential Peirce. Selected Philosophical Writings, vol. 2 (1893–1913), pp. 434–450. Indiana University Press, Bloomington (1998)

10. Dewey, J.: The Later Works, 1925–1953, vol. 1: 1925. Experience and Nature. Southern Illinois University Press, Carbondale (1981)
11. Dewey, J.: The development of American pragmatism. In: Hickman, L.A., Alexander, Th.M. (eds.) The Essential Dewey, vol. 1: Pragmatism, Education, Democracy, pp. 3–13. Indiana University Press, Bloomington (1998)
12. Dewey, J.: The Middle Works, 1899–1924, vol. 14: 1922. Human Nature and Conduct. Southern Illinois University Press, Carbondale (1983)
13. Dewey, J.: Reconstruction in Philosophy. Enlarged edition. Mentor Books, New York (1950)
14. Dewey, J.: The influence of Darwinism on philosophy. In: Hickman, L.A., Alexander, Th.M. (eds.) The Essential Dewey, vol. 1: Pragmatism, Education, Democracy, pp. 39–45. Indiana University Press, Bloomington (1998)
15. Dewey, J.: Does reality possess practical character? In: Hickman, L.A., Alexander, Th.M. (eds.) The Essential Dewey, vol. 1: Pragmatism, Education, Democracy, pp. 124–133. Indiana University Press, Bloomington (1998)
16. Dewey, J.: The Later Works, 1925–1953, vol. 4: 1929. The Quest for Certainty. Southern Illinois University Press, Carbondale (1984)
17. Dewey, J.: A Common Faith. Yale University Press, New Haven (1934)
18. Dewey, J.: Evolution and ethics. In: Hickman, L.A., Alexander, Th.M. (eds.) The Essential Dewey, vol. 2: Ethics, Logic, Psychology, pp. 225–235. Indiana University Press, Bloomington (1998)
19. von Weizsäcker, C.-F.: The Unity of Nature. Farrar Straus Giroux, New York (1980)
20. Küppers, B.-O.: Die Strukturwissenschaften als Bindeglied zwischen Natur- und Geisteswissenschaften. In: Küppers, B.-O. (ed.) Die Einheit der Wirklichkeit. Zum Wissenschaftsverständnis der Gegenwart, pp. 89–105. Fink, München (2000)
21. Artmann, S.: Historische Epistemologie der Strukturwissenschaften. Fink, München (2010)
22. Dewey, J.: The reflex arc concept in psychology. In: Hickman, L.A., Alexander, Th.M. (eds.) The Essential Dewey, vol. 2: Ethics, Logic, Psychology, pp. 3–10. Indiana University Press, Bloomington (1998)
23. Küppers, B.-O.: Nur Wissen kann Wissen beherrschen. Macht und Verantwortung der Wissenschaft. Fackelträger, Köln (2008)
24. Simon, H.A.: Cognitive science: the newest science of the artificial. Cognit. Sci. **4**, 33–46 (1980)

System Surfaces: There Is Never Just Only One Structure

Klaus Kornwachs

Ich weiß ein allgewaltiges Wort auf Meilen hört's ein Tauber.
Es wirkt geschäftig fort und fort mit unbegriffnem Zauber ist
nirgends und ist überall bald lästig, bald bequem. Es passt auf
ein und jeden Fall: das Wort - es heißt System.[1]

Franz Grillparzer (1791–1872)

Abstract When making modes, it is a constitutive question, how a system can be described and where the borderlines between a system and its environment can be drawn. The epistemic position of system realism (there are systems as ontic entities) and system nominalism (systems are only descriptions) leads to the concept of natural surfaces on the one side and to the authorship of defined borders on the other side. A more mediated position, the system descriptionism, is unfolded here: it is shown that changing objects or processes, which are described as systems, require new system theoretical concepts. As one of them, the notion of non-classical system is introduced, and a classification thereof is proposed.

1 Introduction: Concepts of System and Structure

If we conceptualize a process or an object *as* a system—are we then talking about a real objector about a construct? Obviously there is a basic tension between a system concept that is conceived as a methodological tool and a system concept that

---

[1]Franz Grillparzer (1791–1872), Austrian Poet, cf. [32, GG VI 316, p. 463].

K. Kornwachs (✉)
Brandenburgische Technische Universität Cottbus, Cottbus, Germany
e-mail: Klaus@Kornwachs.de

represents a system as an object with a "natural" surface or as a genuine entity within the world. The aim of this paper is to focus on the concept of surface of a system. Due to the tension mentioned above the concept of surface remains ambiguous. This becomes clear when changing systems are observed. In order to avoid problems with any misleading system ontology, the position of descriptionism is introduced.

A first step[2] to do so is to look for some historic predecessors of system ontology.[3] The first system concepts have included systems as a means of description with the purpose "to order," "to induce order," "to detect order," or "to observe order" within a framework of what we call reality. This relation between system concept and the notion of order seems to be very old. One may find it in pre-Socratic fragments as well as in medieval writings.[4]

The second concept includes system as an object or an entity with genuine properties like closeness, stability, growth. Its existence that encompasses dynamics as becoming and vanishing, generation and corruption.[5] Both concepts have a long history within the ongoing development of natural sciences and particularly in philosophy as well as in philosophy of nature.

The roots of this development may allow suspecting that system thinking was in earlier time a good candidate for a comprehensive theory of everything (respective explaining everything).

1.1 Historical Remarks

Only a few examples may illustrate this point of view. In attic philosophy for Aristotle system was a whole entity. A tragedy or the state served him as an example.[6] Here we have two meanings: $\sigma\acute{v}\sigma\tau\eta\mu\alpha$ expresses an active putting together like a combination of different and equal things, whereas the Greek verb $\sigma\upsilon\nu\acute{\iota}\sigma\tau\eta\mu\iota$ describes the action of unifying, ordering, putting together. In later Stoa philosophy system was an overall concept for things used in practical sciences.

G. W. Leibniz intended with his *Monadologie* to give an elementary theory of everything, every object, and every process [78]. Originally conceptualized to answer the question of substances in ontology together with a *Theodizee* (how a good God can allow or tolerate the evil within the world), all monads form a

---

[2]The following section has been adopted and modified from a former paper, cf. [68].

[3]Here we do not use the term "ontology" as it is used in the realm of computer science like in artificial intelligence, data base technology, and others (universe of discourse). We refer to the original term ontology as it is used in philosophy. Ontology is the teaching about being (*Sein*), existing things (*Seiendes*), and modalities thereof.

[4]Cf. for instance Heraklit: Über die Natur. Fragment 1 and 2 [15, p. 77] or Fragments 10–13 [15, p. 131f.]. For medieval thinkers take for an example Heinrich Gent [29, p. 564 ff.]. Cf. also Hübner's article about order (*Ordnung*) [50, pp. 1254–1279].

[5]Like the Aristotelian book (1965) with the headline *de generatione et corruptione* [4].

[6]Cf. [3, Poetry, Book 18, 1456a11] or in [5, Nicomachian Ethics, Book 9, 1168b32].

whole of being, made stable by a god who designed the best of all possible worlds. This pre-stabilized harmony was an order that assigned the characteristic of an existing system to the world. Moreover the pre-designed system structures could be explained as the laws of the whole system. Each monad is different from each other one such that they present a kind of subsystems and G. W. Leibniz introduced a certain kind of hierarchy between the different sorts of monads. Nevertheless no monad could substitute another one or be removed from the best of all possible worlds.

The reason why this example is stretched here may be given by the fact that in such a kind of system conception the real existing world or objects have the character of a system because they have been designed or created as systems. Thus systems are real things and they can be distinguished from each other and all systems form a giant supersystem that rules everything. The system characteristic is corresponding to the design that presents the best thinkable order. Here, order and existence coincide: order is not a perspective or result of an evaluation, but a genuine property of existing things and of the world. This concept may serve as a paradigm for what we call classical ontology.

R. Descartes changed the number of substances, simply reducing it to two: *res extensa* and *res cogitans* [21]. Speaking about systems, he referred to a complete and ordered set of knowledge. To know something about an object means to know its shape (Gestalt), its structure, the border between object and environment, its extension, its existence, and its individuality.[7] Descartes tried to establish philosophy as an autonomous building of knowledge. Thus in his "Principles of Philosophy" [21] he aimed to certify, to justify, and to found philosophy as a system by itself. For Descartes any systematization means to proceed in an ordering manner. Then a system is the result of such a process. Firstly the laws of system are the laws of system description. But this description can be only successful if there are coincidences between the laws, the methods of description, and the properties of the object described. Due to the fact that an object has the property to be a system, it can be described as a system. Therefore a description, using the mathematical character of nature, can be successful, because it applies the principles of mathematics.

Not the things are systems, but our concepts about them. Prepared by Descartes, this is a dramatic change because it de-ontologizes the system concept and shifts the concept of order from ontology to description. Later, the system of Pure Reason has been conceptualized by Kant [53] as an epistemological prerequisite that foregoes all possible human experience like Geometry, Aristotelian Syllogism, or even Newton Mechanics (cf. also [52]). The system character of this precondition for the possibilities of empirical experience lies in the order which we believe to find by experiences. Subsequently we believe to find this order in nature and in the nature of things as well (Fig. 1).

G. W. Hegel proposed a system of historic development ruled by dialectical laws corresponding to his System of Philosophy [38, pp. 11–151]. Exactly here we have

---

[7]Cf. a comprehensive summary by [112, p. 1463].

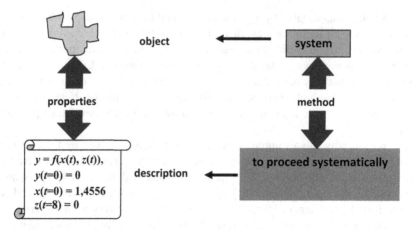

Fig. 1 Corresponding coincidence of system description, method, and the characteristic properties of the described object, according to Descartes

a new development in system ontology. Whereas the order of Descartes' or Kant's system is an order in thought where reason is a matter of concepts about things (on a descriptive level), G. W. Hegel maintained that reason comes to existence, and whatever is existing would be reasonable (cf. [39, p. 14 (XIX)]). Thus Hegel re-ontologized the system concept and connected again the concepts of order (as a metaphor for reason) and existence. From G. W. Hegel to N. Luhmann there is an obvious red line: systems really exist, they come into existence by their own dynamic and intrinsic laws [83]. Under this regime N. Luhmann described social systems as genuine existing, autonomous systems with observable properties and recognizable inner structure, regularities and laws of their inner nature.

With this short sketch[8] we are able to distinguish between two basic system concepts and the following thinkers and authors may be grouped along this classification. The conceptual order we apply to describe real or intelligible things allows us to use system laws (of formal, logical, and mathematical provenience) to make forecasts, explanation, rules for control, and so on of the things described by these concepts. Thus for Frege [27] only Logical Systems, i.e., only conceptual frames, were considered as systems. H. Poincaré put forward a more conventionalist variant: systems as an agreed conceptual framework to arrange the manifold of the recognized issues (cf. [104]). The late M. Heidegger conceived system theory (cybernetics) as a Philosophy of Science[9] and even J. Habermas regarded systems

---

[8]For a more extensive presentation of the history of system concept cf. [35, 102, 104].

[9]Heidegger [40] pointed out that a system is encompassing all what can be deduced. It *induces structures and changes all knowledge of science, it saves it (the knowledge; the author) and brings it to communication; it may be conceptualized as the totality of grounding relations, in which the matter of science (as objects, germ.: "Gegenstände"; the author) is represented with respect to its basis, and where they can be understood conceptually* [40, p. 56] (transl. by the author). This has been also explicitly pronounced by Heidegger in his famous SPIEGEL-Gespräch (1976) with

(i.e., models) as a result gained by systematic method of the researchers (cf. [33, 34]). H. Lenk has analyzed the system concept itself by means of Philosophy of Science and discovered that system is a perspectivistic concept [79]. G. Ropohl developed further the constructivist approach that the researchers' perspective is constitutive for understanding systems [100]. E. Lazlo has prolonged this idea in order to try to build a Model of Epistemology [77].

We could also distinguish—as a working hypothesis—the two positions as system realism (systems do exist in an ontological modus of reality) and system nominalism (systems are only descriptions of a set of objects, processes, properties as a section of (hypothetical) reality by formal means).

As a result of this historical development we have today a bundle of partial disciplines, weakly connected only by mathematics. Nevertheless, a meta-theoretical interest for the nature of these connections seems to be lacking. One could ask whether these disciplines[10] may form a backbone for a new structural science. As a matter of fact the concept of structure is used in a digressive manner. One definition refers to the relation between elements or subsystems. Another concept of structure relates to classes of functions appearing within the dynamic state transition equations. A further concept of structure is given by classifying solutions of dynamic equations as types of behavior of a system. Sometimes a characterization of surfaces of objects is also called a structure, and in physics one is used to speak about symmetries in terms of structures.

1.2 Basic Definitions

It may be useful to sum up here shortly some paradigms of classical system descriptions. "Classical" means here that the classical mechanics according to the mathematical basis given by Hamilton and others serves more or less as basis. One can define state space, where the dimension is given by the number of independent variables or observables. The dynamic is given by a mapping of the influencing or independent variables (= input) on the state variables and the influenced resp. dependent variables (= output). The separation between such classes of variables is not self-understanding, for George Klir this distinctions lead to controlled systems [55, p. 282]. The dynamic of a system can also be expressed by state transition

---

Augstein [42]: *What will come after Philosophy? Cybernetics. But this will not be Philosophy anymore.* (transl. by the author).

[10]Like cybernetics or control theory [109], theory of automaton [88, 91], theory of syntax [20], game theory [90], theory of communication [57, 58, 103], and general system theory [9]. All these theories deal with state space approaches, at least homomorphism between syntax theory and theory of automaton can be shown, discrete cybernetics are automata, and continuous automata can be modeled as control circuits. Game theory can be modeled as automata, and the theory of communication can be formulated either as statistics of stochastic signal processes or even as noisy automata (i.e., communication channels, receivers, and emitters). These theories may considered to be classical theories (see Sect. 4).

Fig. 2 Formal system
description (For definitions
see text)

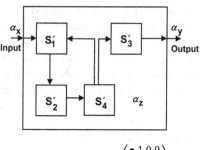

$$\text{Structure } \pi: \quad \kappa_{ij} = \begin{pmatrix} \bullet & 1 & 0 & 0 \\ 0 & \bullet & 1 & 0 \\ 0 & 0 & \bullet & 0 \\ 1 & 0 & 1 & \bullet \end{pmatrix}$$

equations in analogy to the equations of classical mechanics. They show the motion of the system within its n dimensional state space along its trajectories. Using all that, the control theory has developed the concept of behavior of a system: it is represented by the dynamic which depends on state transition laws and boundary conditions. If laws and boundary conditions are known, the properties of observability, stability, controllability, and reachability can be investigated [93]. In this concept, a structure is represented by the connections between the elements or subsystems. In classical mechanics these connections are given by forces, in system theory this is generalized to influences or interactions. The decisive dependence between behavior and structure is given by the postulate that the overall behavior of a system can be calculated in principle,[11] if the structure and the basis behavior of the elements are known.

Formally written, the system can be represented by a graph and the structure can be expressed like in the following example (Fig. 2).[12]

According to this a system can be defined by a quintuple

$$S = \{\alpha, \phi, \sigma, \pi, T\}. \tag{1}$$

Let the variables of the system be given by

$$\alpha = \{\alpha_x, \alpha_y, \alpha_z, t\} \tag{2}$$

with $\alpha_x \in \{X\}$ as variable of the input, $\alpha_y \in \{Y\}$ as variable of the output, $\alpha_z \in \{Z\}$ as variable of the states, and $t \in \{T\}$ as time basis. The state transition function

---

[11]In this rigidity the statements hold exactly only for linear type behavior of the elementary behaviors.

[12]The structure, shown in Fig. 2, serves as an example.

(= dynamic of the system) δ and the output function λ represent the overall behavior of the system, written as

$$\phi = \{\delta, \lambda\} \quad \text{with} \quad \delta : \alpha_x \otimes \alpha_z \rightarrow \alpha_z \quad \text{and} \quad \lambda : \alpha_x \otimes \alpha_z \rightarrow \alpha_y. \qquad (3)$$

Subsystems (or elements, conceptualized as subsystems not to be analyzed more in detail) are connected with each other by the structure

$$\pi = \{\kappa_{ij}\} \quad \text{with} \quad \kappa_{ii} = 1, \qquad (4)$$

and the set of subsystems

$$\sigma = \{S'_S\} \quad \text{with} \quad S' = \{\alpha_{S'}, \phi_{S'}, \sigma_{S'}, \pi_{S'}, T\}. \qquad (5)$$

Iterations are possible down to the wanted level [86]. Nevertheless some presuppositions have to make for this quite simple framework: the structure should be constant, i.e.,

$$\pi = \{\kappa_{ij}\} = \text{const} \qquad (6)$$

as well as

$$\alpha = \{\alpha_x, \alpha_y, \alpha_z, t\} \qquad (7)$$

must be stable during the observation or working time of the system. As a consequence the dynamic

$$\phi = \{\delta, \lambda\} \qquad (8)$$

doesn't change its characteristics. If we simplify the overall behavior with the output function λ as identical with

$$\alpha_z \rightarrow \alpha_y, \qquad (9)$$

then we can write the local behavior ϕ_i of subsystem (or element) indexed with i as

$$y_i = \phi(z_i, x_i) \qquad (10)$$

with

$$x_i(t) \in \{\alpha_x\}, z_i(t) \in \{\alpha_z\}, y_i(t) \in \{\alpha_y\}, z_i(t) = y_i(t). \qquad (11)$$

If the subsystems are connected due to a given structure $\pi = \{\kappa_{ij}\}$ more than one x_j is influencing the ith subsystem. Writing the local dynamics ϕ_i as linear operators, we can calculate the overall behavior

$$\Phi = ||\phi_i \delta_{ij}||, \qquad (12)$$

$$K = \{\kappa_{ij}\} \qquad (13)$$

with δ as the Kronnecker symbol by solving the set of linear equations

$$\Phi X = Y - \Phi K Y, \qquad (14)$$

Fig. 3 System border and membership of elements: e_i has no relations with other elements of the system, therefore it doesn't belong to the system. The *arrows* indicate the relations between the systems

Border between system and environment

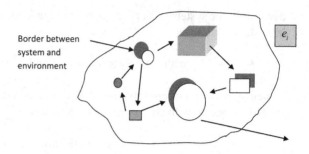

with the input vector

$$X = \begin{pmatrix} \vdots \\ x_i \\ \vdots \end{pmatrix}, i = 1, \ldots, n \tag{15}$$

and output vector

$$Y = \begin{pmatrix} \vdots \\ y_j \\ \vdots \end{pmatrix}, j = 1, \ldots, m. \tag{16}$$

The structure here is defined iteratively with $\pi = \{\kappa_{ij}\}$ according carriers of the interaction between the elements or subsystems.

A surface of a system conceptualizes as above can be defined in three ways: first, one could take the input vector X as the set of all independent variables. The second idea is to take all input and output variables X and Y as an expression for the border between inside and outside of a system. The third definition is preferred here: the structure π defines which element is belonging to the system and which element does not. For a not connected element i the properties $\kappa_{ij} = 0$ for all i and $\kappa_{ji} = 0$ for all j hold; i.e., e_i doesn't belong to the system (Fig. 3).

2 System Surfaces: Objects and Their "Authors"

2.1 "Obvious" Surfaces

After this more formal side of the system concept the question may arise how and by whom the borderline between system and environment (= nonsystem) is drawn. If we take an object we constitute a kind of a surface by observation. Usually we can separate objects from each other, in most cases we can divide objects in several parts as sub-objects, i.e., we can make a cut between object and environment by an action we could call preparation. In Philosophy of Technology this is well known as the transition from *presented-at-hand* (as an *objet trouvé*) to the *readiness-to-hand* (the tool).[13] This constitutes an inside and an outside. The same can be done conceptually—we may describe a system from an outside perspective, as done for example by behaviorism and its concept of a black box, or a rather inside perspective, using analytical methods to find out structures and elements. Both views can be justified very well, an extended discussion can be found in terms of exo- and endo-view [6].

Using Descartes notation of extension one could generalize the concept of system surface to four physical dimensions. A four-dimensional extension defines the topological spatial limitation of the extension as well as the time limitation of existence of an object, i.e., the very beginning and the very end. There are border areas of this concept: the atoms of Democrit are not dividable and therefore they have no beginning and no end in time.[14] Fractal objects like the Menger Sponge or Cantor's Dust are another example—potentially they have an infinite surface. Looking for very large systems in cosmology or even in large technical systems, the concept of subsystems and elements is only meaningful if they can be "reached" in terms of interaction by the whole system in an appropriate time.

For dynamic black box systems the set of independent, i.e., influencing variables, and dependent, influenced variables constitutes the surface of a system, together with the constraints (initial and boundary conditions). The latter is defined or put in practice by the preparation as a precondition of an experiment.

This is the question of how we can observe obviously or seemingly existing surfaces, i.e., how borders between the inner and the outer side of objects and "systems" as objects, processes or ordered set of them, can be described as systems? (Fig. 4 and Table 1)

To solve this problem, we may make a very rough distinction between different types of how a surface could be constituted [68].

System surfaces, recognized by observation, behave like simple objects. There is a clear cut borderline between the box (as a synonym for an object) and its environment. The simple square in Fig. 5 may symbolize this isolation from a surrounding environment. The observation can be verified by stating that a selected property (value of a variable) is changing twice if one passes the borderline, symbolized by the arrow (like color, etc.).

---

[13]Cf. for this distinction at [41, Sect. 15, p. 69 ff.] between a present object, found by man, i.e., *Vorhandenes = presented-at-hand* (as an *objet trouvé*), and a tool as *Zuhandenes = the readiness-to-hand*. For the translation of this concept by J.M. Macquarrie and E. Robinson cf. [41, p. 98 f.].

[14]Cf. Democrit in [15] pp. 396–405 (fragments 68 A) and p. 418 (fragment 68 A 71). Notation according [22], Vol. 2, p. 41 ff. and p. 95 ff. respectively.

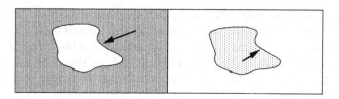

Fig. 4 Outside (*left*) and inside (*right*) aspect of an information transfer through a system surface

Table 1 Outside and inside (cf. [66])

From outside to inside	From inside to outside
Surface: where the system begins	Surface: where the system ends
Outer surface gives information (contains information) about the system	Inner surface contains information about the environment
Information (i.e., symbols[a]) about the system must be written from inside on the outer surface, so that the neighbored system can read it	Information (symbols) are written from outside on the inner surface, so that the system can read it
Symbols refer to inner properties	Symbols refer to outer properties
Environment interacts with system	System interacts with the environment
Changing outer surface = changed behavior or structure *for* outside (observer)	Changing inner surface = changed behavior or structure of the environment (world) *for* inside
Overall behavior	Inner structure, elements

[a] Symbols are signs which are already interpreted in a data model. They represent the elementary units of information

Fig. 5 Isolated "object"

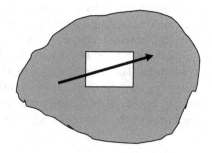

The constitution of a system surface by experience is related to the difference between "inside" and outside. Here not a variable changes its value passing the borderline, but the system is constituted by the experiences one can make with "it" as a quasi-object or process. The experience can be expressed in terms like behavior or input-output relations, but this constitution is an implicit one.

The ingoing arrows in Fig. 6 may be interpreted as independent influences (variables), the outgoing arrows as the resulting or dependent variables. The selection of the set of variables constitutes the functional surface of the system (Fig. 6).

Fig. 6 Implicit surface of a
system by input and output
variables

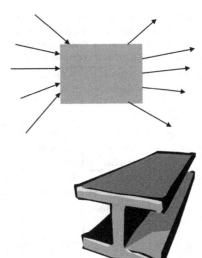

Fig. 7 Object/system with an
"obvious" inside/outside
surface

The constitution of the system surface by description (Endo–Exo) presupposes both the forgoing procedures: variables and properties are selected and the embedding of the system within a supersystem and the division into subsystems are already a result of a system analysis. This is already a system description. At this point we may distinguish between an inner surface (the surface observed from inside) and an outer surface (the surface observed from outside), suggested by Fig. 7.

At this very point one could argue that there are ontological system surfaces, suggested on an ontic level. On the left side of Table 2 such kinds of system surfaces are listed.

A possible answer to this serious objection could be: it is not the system description that induces a surface or the borderbetween inner and outer side, but we choose a system description which mostly runs along the ontological given surfaces of already separated objects or objects separated by us. The reason for that is quite simple: we try to get descriptions that are easy and simple in order to proceed further with a minimum of effort. It seems to be that one hopes that surfaces in system theory that are designed according to ontological surfaces are simpler than other ones. Indeed, in an overwhelming number of examples this is the case. But we should not be worried about counterexamples.

Thus the borderlines between *exo* and *endo* can be drawn with more advantage (i.e., the mathematics of the description becomes simpler), if one takes the choice with respect to the aimed function of the model instead of taking the surface that is given by phenomenology (cf. also [66]).

The advantage of adaptedness of a system choice (i.e., model building decisions) depends extremely on the use of the system description the author has in mind (explanation, forecast, simulation, control, etc.).

Table 2 *Left four rows*: surfaces of different system types. According to the different surfaces the transfer of material, information, and energy flow may be observed. The *two rows at right*: by interest of authors of the system and by changing constraints, the definition of the initial surface of the system in account is extended

System type	Inner surface	Outer surface	Transfer by	Extension of the given surface by	Changed perspective are given by:
EDP, computer	Layer behind I/O-elements, compiler	Screen, display, keyboard, sensors	Signals	Ubiquitous computing[a]	Developer of a technical "system"
Biology	Organ surface, membranes	Epider-, mis skin	Metabolism, diffusion, osmotic process	Transition from individual organism to ecosystem[b]	Changing boundary conditions, implementation of inner binding interactions
Physics	ψ-function	Observable	Measurement	Transition from particles to an ensemble	Free choice by model building
Technology (Car)	Motor surface	Instrument panel	Sensors, actors (force transmission)	Device $\Rightarrow$ organizational closure	Functional interest, e.g., externalize a given function
Organization	Procedure	Legality	Decisions	Super-organization	Doubts in competence and legitimacy
Networks, nets	Connectivity	Devices	Information	Communication strategy, connection policies	Networks in competition
General systems	State $\rightarrow$ signs	Signals $\rightarrow$ information	δ-, λ-function	Embedded system $\Rightarrow$ embedding system	Model building and control interest

[a] Cf. for UC in [69]

[b] Another extension could be considered within this context: the transition from an individual entity to a swarm; cf. [12]

2.2 Epistemic Issues

In a further step we would like to contrast two epistemological points of view, the realism and the descriptionism.

In realism systems are existing as such, and therefore there are also objectively given laws of such systems. Taking natural or artificial systems like electronic devices, physical objects, and technical devices, they all obey to natural, i.e., physical laws. Even within the realm of society and organization, systems like societies, communities, or institutions do exist in a plain ontological manner [83].

In the context of descriptionism, systems do not exist as an outer object (Gegen-stand), but systems are descriptions of selected facts, properties, and processes. There are laws according to the circumstance that the facts, properties, and processes can be understood, and there are laws of the system description, dependent on the chosen language of description, like mathematics, grammar, logic, control theory, or cybernetics. Thus, natural or artificial systems like electronic devices, physical objects, and technological devices are described as a connected and wired net of electronic elements, as physical objects or as techno-organizational systems. They are modeled for example with the help of the equation in Chap. 1.2 or by a state space approach with coupled equations or by event oriented simulation models. Even societies, institutions, or organizations are modeled in analogy to the known laws in biology, physics, or control theory.

It is clear that descriptionism makes no statement about reality and ontology. The modeling approaches use knowledge about the field in which the *Gegenstand* should be modeled as a system. To use physical knowledge to find appropriate state space equations or to use cybernetic knowledge to find the appropriate control equation is not yet a statement that the *Gegenstand* obeys physical laws, but a modeling strategy using obvious or presumed analogies. In other word: who is writing a model of a *Gegenstand* is the author of the system. Each system has an author [59].

3 There Are no Systems

3.1 The Position of Descriptionism

Let us unfold the position of descriptionism more in detail. A good test is the question: what is an open system?

If we had, according to Kant, a pre-theory we could observe such systems—openness and systemicity (as the property to be able to be a system) would be the necessary forms of *Anschauung* (i.e., forgoing all experiences) in order to be able to make system experiences.

A potential answer we could get from thermodynamics: if one is able to describe the constituting parts of a systems in terms of non-equilibrium thermodynamics (NEThD), then it can be calculated like a NEThD. Some characteristics can be explained like the difference between the inside and outside entropy. Other traits like reduplication or growing and shrinking cannot be explained in terms of NEThD.

The approach of constructivism, put forward by Heinz von Foerster and others [25, 26], states that any "thing" we can observe is calculated by the brain, based on sensory data and already available models or model like imaginations. The question of true or false is not relevant—the justification and the validation of the model are at least given by its successful application with respect to the respective survival function for the "user." Nothing is said about the truth of the model, only about its successfulness.

The descriptionism tries to modify the strict position of constructivism: each attempt to model a phenomena or a *Gegenstand* is constrained by the purpose of this modeling like control, explanation, forecast, simulation, or technical production. Each model process draws a borderline between the parts belonging to the system and other entities not belonging to it. Such borderline constitutes a surface of the system. This surface can be conceptualized as an expression of the interests of the model builder about his *Gegenstand*. In this case there is no another one. Thus each system as system description has an author who pursues his interests when dealing with his system description.

This is obvious in system theory. Within the context of Philosophy of Science this authorship should be recognized as a basic trait of scientific modeling. In other words: system theory is a certain kind of Philosophy of Science. Any *general system theory* is a tool for the describing, modeling, and mathematizing interesting fields of objects.

Selected fields of objects which we describe as systems seem to behave like separate entities (say objects with surfaces). Normally we talk about them as they would be existing systems with an ontological status. Nevertheless we should take into account that it is our choice of description types (or say model types) which generates the condition under which a system description is adequate with respect to our interest in the field of objects. One could accept reality, i.e., for circumstances independent of us, as a hypothesis, but without a modeling scheme we cannot make any observational or descriptive statements. According to Kant each observation presupposes basic concepts, stemming from a preliminary description resp. pre-theory. Thus system laws are no natural laws but laws of description. Laws of description are always formal laws.

We conclude that a system description is not only given mathematically a priori, so to say as a mere specification of set theory.[15] There is of course an actual relation between the level of description and the level of objects, since each modeling uses knowledge about the field of objects.

If we conceive the inputs and outputs of an observable system in terms of information, i.e., if we model the input and output variables as representing information acting between the objects, conceptualized as systems, then we can correlate changes within the field of objects with changes of system descriptions.[16]

---

[15]The set theoretic definition of a system is given according to [86, p. 254 ff.]. A general system is a relation on abstract sets $S \in X\{V_i; i \in I\}$ with X as the Cartesian product. I is the index set. If I is finite, it is $S \in V_1 \times V_2 \ldots V_n$; the components of the relation V_i are called objects of systems. They can be divided into the input variables $X \in \{V_i; i \in I_X\}$ and the output variables $Y \in \{V_i; i \in I_y\}$ such that an input/output system can be written as $S \subset \{X \times Y\}$. The definitions in Chap. 1.2 can be regarded as specifications of this concept.

[16]Let us presuppose that some modifications in the mathematical apparatus of system theory are possible. These modifications are discussed in [87, 108], and [61] with respect to numerous applications and disciplines.

Due to this, system theory loses its pure mathematical character and becomes a tool and part of an empirical science. As such it is subject of a transcendental-pragmatic procedure of foundation,[17] which conceives system design and modeling as a question of practical philosophy as well as the use of system descriptions in order to control the relevant field of interest.

3.2 Relation Between the Object Level and Level of Description[18]

> On the object level of a scientific discipline there
> should be no any a priori statements.[19]

One could argue that there is a contradiction between the two positions, developed above: on the one hand, system is a construction using mathematical tools. On the other hand system theory is an expression and part of empirical science.

As far as we know the method of a description and its elementary units, the formula of statements, similar to mathematics and logic, cannot be found by empirical data. They are analytical judgments a priori according to Kant.[20] This was at least the saying of the nineteenth and beginning twentieth century: as far

---

[17]Structurally seen, a method to found a system of rules, i.e., for moral philosophy, in a pragmatic-transcendental way (acc. to [2]) is coextensive with a method of foundation for rules of the use of a system. More in detail: any pragmatic interpretation of scientific law leads only to a rule of action, if the validity of the law and the acceptance for the necessity of existence thereof are unanimously accepted as an a priori in the realm of a dispute procedure. If, when designing a system model, validity and acceptance cannot be denied without contradictions, then one has a constitutive presupposition for the design itself.

[18]For this section cf. also [63, Chap. I.4.].

[19]Oral communication by Walter v. Lucadou, 1974.

[20]This is not very original, but it used to be forgotten quite often. The relevant passage at Kant in Critique of Pure Reason: *Philosophical cognition is the cognition of reason by means of conceptions; mathematical cognition is cognition by means of the construction of conceptions. The construction of a conception is the presentation a priori of the intuition which corresponds to the conception. For this purpose a nonempirical intuition is a requisite, which, as an intuition, is an individual object; while, as the construction of a conception (a general representation), it must be seen to be universally valid for all the possible intuitions which rank under that conception.* ["Die philosophische Erkenntnis ist die Vernunfterkenntnis aus Begriffen, die mathematische aus der Konstruktion der Begriffe. Einen Begriff konstruieren heißt: die ihm korrespondierende Anschauung a priori darstellen. Zur Konstruktion eines Begriffs wird also eine nicht empirische Anschauung erfordert, ohne folglich, als Anschauung, ein einzelnes Objekt ist, aber nichtsdestoweniger, als die Konstruktion eines Begriffs (einer allgemeinen Vorstellung), Allgemeingültigkeit für alle möglichen Anschauungen, die unter denselben Begriff gehören, in der Vorstellung ausdrücken muss."] (cf. [53, B 741], translated by J. M. D. Meiklejohn; cf. http://www.gutenberg.org/cache/epub/4280/pg4280.html).

as mathematics starts to deal with reality, it is not exact.[21] The calculation of a satellite orbit or of a rocket is, despite the validity of Kepler's and Newton's laws, a comparable extensive procedure done by approximations.[22] The formal exact description of Newtonian law, expressed by a differential equation, is only a necessary condition for its empirically successful application. Mathematics doesn't tell us anything sufficiently about the physical validity of such an expression. The necessity of the presupposed formal validity has nothing to do with the field of objects; it is only a presupposition for handling the formal apparatus of a calculus in order to make rule-based conclusions and proofs.[23]

Therefore we can say that any mathematical formalism of system theory cannot generate or substitute knowledge, even empirical knowledge from the field of objects. In other words: the knowledge we need must come from the field of objects. We can only generate a set of consequences from basic statements about the field of objects when using formal methods.

Observing and collecting empirical data is only possible under the regime of an already available concept. Without stressing further the debate about realism,[24] one could come back to the dictum of Einstein according to whom just theory determines what can be observed.[25] For a quantitative observation a quantitative concept is necessary which has been conceptualized in advance—may be hypothetically, speculatively, or only by intuition. This concept contains implicitly a rule how to operate observation or measurement. The semantic meaning of this concept is given by these rules with respect to the field of objects. A measurement device can be conceived as a system, relating the concept and the measurement. Only within the context of this relation the measurement device can produce an information, i.e., the result of measurement. The meaning thereof is ruled by the concept. Only with the help of system description the measurement result can be interpreted meaningful. One could say: the surface of a system rules the possibilities for observation, measurement, and experiment.

---

[21] Einstein has paraphrased this Kantian position in his famous lecture for the Prussian Academy of Science about "Geometry and Experience": *As long as mathematical propositions refer to reality they are not certain, and as long as they are certain* [well proved], *they don't refer to reality.* Cf. [23, p. 119 f.] (transl. by the author).

[22] See textbooks for orbital calculations [11].

[23] These conditions are not valid, if the concept of derivability and the concept of conclusion (i.e., the concept of truth and provability) don't coincide anymore. This is the case in higher level calculi, proved by Gödel [31]. This leads to a lot of restrictions for system descriptions when dealing with complex systems that must be modeled by such higher level calculi. Cf. [63, Chap. III.2.].

[24] See above; for [17, Part. Sect. 9–11], this was a pseudo-problem of philosophy.

[25] Heisenberg [45, p. 92] reported about an entretien with Albert Einstein in the year 1927. According to that, Einstein said: *From a heuristic point of view it may be useful to remember what we really observe. In principle it is completely wrong to ground a theory on only observable entities. The reality runs the other way around: It's just the theory which decides what you can observe.* (Transl. by the author). Later, Heisenberg adopted this point of view: *We have to remember that what we observe is not nature* [here] *itself, but nature exposed to our method of questioning.* Cf. [44, pp. 58, 41].

The choice of the concept corresponds to the definition of the borderline between the inside and outside of the system. The measurement device is described in terms of systems. The property constituting the measurement device is the ability to produce information ruled by the concept. This property is invariant from its physical realization; it is rather a function.[26]

Since the choice of the concept constitutes instrumentally what will be measured, i.e., by which concept system and information are ruled, this choice also determines inasmuch a system description will lead to empirical consequences. That means that this choice determines whether it is possible to transform the description into a physically realizable, instrumentally mediated function. In general the use of our language neglects this difference and calls the measurement device as part of the field of objects as well as the process to measure as "system."

Indeed one is forced to make a choice on the level of description due to the empirical consequences. The protocol statements (according to [16, p. 232 f.]). containing empirical terms, which can be formulated on the level of objects by the help of observation, can only be semantically valid, if they are interpreted on the level of description.[27] Beyond the level of description of the system object, measurement device, and measurement there are statements possible on formal structures of description–these are mathematical and logical statements with a priori characteristics. A reduction from this meta-level to the level of description or the level of objects is not possible due to reasons discussed in Philosophy of Science.[28] Therefore there are no a priori statements within a field of objects of a system theoretical description.

System theory is a construction if it is operating on a conceptual level, i.e., if it is used for description. Thus system theory expresses a theory of possible description forms with respect to available concepts.

Possible forms of observation are coextensive with possible structures.[29] Structure is an extensional concept, not a concept of observation.[30] System theory could therefore be a good candidate for a theory of possible structures.[31] Thus system theory remains on the level of description. Nevertheless system theory may become

---

[26]The concept of function has not yet made explicit. Nevertheless it is comprehensible in terms of a mathematical function as a mapping; cf. [60].

[27]This is expressed by the so-called Ramsay Theorem; cf. [16, pp. 246–254].

[28]For reductionism see also [82] and [63, Chap. III.4.].

[29]This is due the fact that we are used to design concepts. Adopting the picture of Klix [56, pp. 536–549], it is clear that each observation based science presupposes the possibility of invariants. Invariant properties can be characterized by a structural difference to what one can observe. Each object seems to be higher structured as its environment—this is what an observer states if he speaks about an object. What can be observed potentially from the forms of observation are possible structures of objects. These and only these structures correspond with the possible forms of observation. This is the reason to use the term "coextensive."

[30]This is not a concept which could be interpreted as an operative measurement rule (*Messvorschrift*). To each observable there is a concept, but not vice versa, cf. [16, p. 59ff.].

[31]Here structure has the meaning we ascribed to it in Sect. 1, Fig. 2.

an empirical science by preparation,[32] i.e., by isolating conceptually the system from the environment and by fixing its constraints or by defining the possible inputs and outputs. With this concept we constitute an object as a real existing thing as far it is covered by the concept applied. The behavior of this object can be forecast due to the system description only if the real behavior is represented by information which can be interpreted in terms of the system constituted by the given concept(s). Whereas the quantum theory represents a theory of possible behavior,[33] we could say that system theory represents a theory of possible measurement or observation devices.[34] If this hypothesis is accepted, we could further say that system theory contains the quantum theory as a particular case if it expresses a theory about possible experiences. Then system theory is—under the regime of human epistemic categories—a theory of possible recognizable whole entities, say objects or things, and their structures. Therefore it is more powerful than physics with respect to its ability for constructing.

4   Changing Descriptions of Systems

4.1   New Phenomena Requires New Descriptions

During the twentieth century new phenomena has been discovered which forced to leave the system theoretical paradise with linear relations and normal state space approaches.

Thus in control theory the non-interaction-free observer is well known. Here the change of an observed behavior due to the interaction between system and observer is described. This induces a new structure of state space approach for the description of the observed system. One may find this in textbooks of control theory: it is flabbergasting to discover here dual state spaces usually applied in quantum mechanics.[35] From a phenomenological point of view one has the same effect— the microscopic particles like electrons are so small that their physical behavior

---

[32]The concept of preparation is discussed intensively with respect to quantum mechanical measurement. The preparation of the constraints is an ontic presupposition for the availability of a system and therefore for an experiment. On the level of description we have the definition of the constraints as initial and boundary conditions, together with the dynamics of a system in order to make predictions.

[33]As proposed by Weizsäcker [107, p. 259].

[34]This is no contradiction to the saying above according to which system theory could be a theory of possible structures. A measurement device presupposes an elementary structure, realized in an apparatus that is defined by the concept, which is the base of the observed variable. Thus an observed structure must have a relation to the elementary structure of the measurement device. This relation is contained within a theory about the field of objects in question.

[35]Cf. [93, Chap. 4, particularly Chap. 4.4, pp. 233–271]. In quantum theory Hilbert spaces are applied. When controlling systems non-interaction-free observed, dual spaces like Hilbert spaces

is changed by the observation due to the necessary energetic interaction between observing device (e.g., photon beam) and particle. This leads to consequences like the uncertainty relation of Heisenberg,[36] according to which location and impact (or time and energy) of a particle cannot be measured at the same time with arbitrary exactness.

In control theory, we have as a consequence of an observation not free from interaction that from the measurements of the observer we cannot conclude anymore on system properties like stability, reachability, and controllability or feed-forward.[37]

This property, i.e., not being able anymore to estimate or measure in which state a system definitely may be, we will call here one of the "non-classical" properties. They are well known in quantum theory as in control theory though without regarding them under the same point of view. The scientists in control theory are not familiar with quantum theory and quantum theorists normally don't care about cybernetics and control theory.

The effects found in synergetic, dissipative systems (also known as Open Systems) and chaotic systems (cf. Table 3) were not less astonishing but more popular in public attention. Changing certain control parameters one obtained a transition from stochastic movements of particles to ordered structures. Nevertheless the mathematical background was already well known since Henry Poincaré: he studied certain types of nonlinear differential equations, the characteristic of the manifold of solution thereof changed drastically when altering certain parameters within the equation. Later René Thom and others investigated such transitions, today, known as catastrophe and bifurcation theory [75].

The solutions (timely developments) of chaotic systems show an exponentially growing divergence, depending upon the choice of boundary conditions. In order to understand the behavior of those field of objects, describable by such nonlinear equations one is relied on statistical statements (like in synergetic), or the dynamic

---

are applied, too, e.g., for linear time discrete systems. Even the problem of nondistinguishable state is discussed there.

[36]Cf. Heisenberg's original paper [43]; with respect to this see also the canonical textbook of Landau and Liefschitz [76, vol. 3, Sect. 16, p. 49ff.].

[37]Nevertheless, an estimation of these properties with observers free from interaction is also restricted when dealing with nonlinear systems. For exact definitions see textbooks of control and mathematical system theory. For illustration here some verbal definitions: *Controllabilty* is given, if it is possible, to "move" the system through the state space from a defined start state into a finite state (or area if states) of interest in a finite time by applying an input. *Stability* is the property of the system to remain at or come back to a finite state or area of states asymptotically, invariant from a given input. *Reachability*: The system behavior has an area or points within the state space that can be reached by a given input from an arbitrary starting point in a finite time. Feed-forward (or feed-through): the system output can be controlled to a given degree directly by the input. To this degree the system can be considered as a channel for the input. *Observability* means that one can estimate the starting point of a system by measuring the input and the output dynamics. Cf. [80, 93]. A real theoretically satisfying formalization of the control problem has been given by [8, 51, 97]. For history cf. also [111].

Table 3 Disciplines of a "new system theory"

	Synergetic	Open systems	Chaos theory
Field of objects	Physical systems + analogies thereof	Thermodynamic systems, symbolic systems	Natural and mathematical systems, fractals
Problems	Forecast of behavior	Understanding of behavior	Understanding of structures
Constituting parts of the system	Stochastic, nonlinear differential equations; Master–Slave principle	H-Theorem, non-equilibrium, thermodynamics	Equations parameter Strange and other attractors
Basic questions	Transition from chaos to order	Generation of systems and explanation of structures	Generation and explanation of chaotic behavior
Authors	Haken [36, 37]	Prigogine [98], Meixner [85], Glansdorf [30]	Thom [105] Lorenz [81]

equation can only be represented recursively without analytical solutions. In other words: chaotic system behavior is deterministic, but not suited for a prognosis by the mean of analytical solutions. One is forced to calculate step by step recursively in a time discrete approximation. This means that the calculation is as long as one would observe the system behavior and writing a protocol step by step. The description of the system is not shorter than the list of data of system behavior. This leaded to the concept of complexity in computer science: a system behavior is complex if its mathematical description is as extensive as the protocol of system behavior in time.[38]

If the behavior of system is extremely dependent on the initial conditions, statements about stability, reachability, controllability, and feed-forward are no more possible in advance but only post hoc. This also relates to the dependence between behavior and structure: it is not possible anymore to conclude from the behavior of elements and knowledge of structure to the overall behavior of the whole system.

Therefore it was hard to withstand the seduction to use the new concept of emergence denoting the appearance of new properties not derivable alone from the

---

[38]The picture of incompressibility of an algorithmic description of complex processes has been developed independently by Kolmogorov and Chaitin [18, 19, 58].

knowledge of structure and elementary behavior, unless the new observed behavior doesn't contradict the theoretical description of system in a retrodictive way.[39]

Nevertheless the generation of new structures from chaotic states of motion (within state space), i.e., the generation of attractors,[40] has led to exaggerated pretensions. Particularly when these models have been applied to other fields of interest beside objects of natural sciences or pure abstract mathematical problems, there have been a lot of disappointments and misleading interpretations.

The development lines indicated in Table 4 are using the system-theoretical description in a same non-classical way: normally systems in AI are large computer programs, performing manipulation of data and symbols and operating with data sets which are structured by diverse relations. These operations are ruled by finite class logic, using predicates. Often logic calculi are called systems, too. It has been shown that the application of pure deductive methods in this field is not very successful when trying to generate an answer for a certain problem using knowledge bases.

There are two possibilities to overcome this obstacle: A so-called default logic has been developed, with which one can formalize general rules and exceptions and derive reliable, i.e., logical true statements from them (e.g., all birds are flying beside Kiwis). Another way is not very pleasant: some inference machines as a part of so-called expert systems, performing conclusions from statements contained in a knowledge base, include also a syllogism called abduction.[41] This syllogism is logically wrong, but applied very often in every day arguing. It serves as a plausibility argument and it is also used in technical sciences, supporting to find functional hypotheses and technical means (cf. [49], especially [28, p. 280 ff.]). This leads to an uncertainty, if one considers the reliability, i.e., the derivability of those consequential statements as answers to questions posed. Therefore one can only talk about an uncertain plausibility of the results of expert system procedures.

Also divergent from the classical concept of system the disciplines, known under the name cognitive science, use the term "system." Herein included are robotics, processing sensory data in order to produce higher level patterns of behavior, as well as neuro-physiologically oriented models of neuronal nets, simulated by physical models (like spin glasses). It became clear that the usual concept of programming (i.e., to write an algorithm) cannot be applied directly in neuronal nets, since the timely development of the net states cannot be predicted being in the learning phase. Sometimes the concept of emergence is used. It is possible to find models that make understandable the appearance of attractors, but time and location cannot

---

[39] A critique of the concept of emergence is discussed in [67] and [70].

[40] Attractors are points or finite areas within the state space, in which the system moves and remains asymptotically. Therefore systems with attractors are stable, but not necessarily controllable, since stability and reachability is only a necessary but not sufficient condition for controllability. See also textbooks about control and system theory.

[41] In contrast to the modus ponens ($[\forall x(A(x) \rightarrow B(x)) \wedge A(e)] \rightarrow B(e)$), the abduction concludes from the particular to the general: ($[\forall x(A(x) \rightarrow B(x)) \wedge B(e)] \rightarrow A(e)$). The proposition logical expression ($[(a \rightarrow b) \wedge b] \rightarrow a$) is falsifiable, i.e., not always true.

be predicted. There is no wonder that the concept of emergence and the non-classical phenomenology has offered a seducing terminology in order to explain badly understood phenomena[42] in social or psychical systems. Therefore we would exclude here the so-called sociological system theories after Niclas Luhmann and other theories of action, since they actually use the system theoretical vocabulary in a metaphorical purpose only. This doesn't lead to real models which could be understood or criticized in a formal, mathematical, or system theoretical modus (according to Tables 3 and 4 both rows left). In the realm of the scientific community in Humanities and Social Sciences the term "system theory" is nearly exclusively used in terms of Luhmann Theory; this development branch has been also included in Table 4.

A presupposition for a classical system definition is the robustness of a system—its structure, behavior, and surface should be constant during the observation or use (like simulation or control processes).

All this new phenomenon share a common ground. The non-interaction-free observer with its alteration of behavior observed forces a new state space structure for description. In chaotic systems the dynamic equation have no analytic solutions, only recursive procedures. The behavior depends extremely upon initial conditions; nothing can be said about stability. New structures are appearing from unordered relations, e.g., the generation of attractors, and we cannot conclude anymore from structure and elements to the overall behavior. All these changes can be related either to the change which we observe in contrast to what we expect (forecast and protocol) or to a change of the frame of observation, i.e., we have changed our view on the system and therefore the surface of the system. The changes appearing to us within the field of objects are due to the change of perspective on the level of description like the change of drawing the borderline between system and environment, a change in constraints and preparation or an altered embedding into the realm of other co-systems.

As an example a self-changing system within the realm of endo-view may serve: it can be described within the realm of exo-view as a mechanical problem (see Fig. 8).

As a first approach we could describe the change of concept: a classical system has a constant structure

$$\pi = \{\kappa_{ij}\} = \text{const.} \tag{17}$$

and the dimensions of the control

$$\alpha = \{\alpha_x, \alpha_y, \alpha_z, t\} \tag{18}$$

remain timely constant. Moreover the dynamics $\phi = \{\delta, \lambda\}$ doesn't change its characteristics.

---

[42]For badly understood phenomena one is not successful to reduce them to law-like principles within the realm of a theory.

Table 4 Disciplines of new system theory (since ca. 1970)

	Knowledge technology	Cognitive systems	Cognitive systems theory of action
Field of objects	Expert-systems, knowledge representation, Artificial Intelligence	Neuronal nets, robotics	Social relations, behavior, acting
Problems	Knowledge processing	Learning, understanding, recognizing, deciding	Understanding processes and structures
Constituting parts of the system	Knowledge base, inference-structures, logic calculus	Neuronal nets, learning phase, feasibility phase, Spinglass—Models	Description in ordinary language using system theoretic concepts
		$\sum \leq \vartheta$	
Basic questions	Formalization of knowledge, simulation of cognitive processes	Understanding and designing of cognitive processes	Understanding of social systems
Authors	Winston [110] Newell, Simon [92] Weizenbaum [106]	McCulloch [84] Palm [94]	Parson [95] Luhmann [83]

Looking on fields of objects in biology with generation, annihilation, growing, and decaying, one is forced to adapt the system description in leaving the classical concept. Now the structure

$$\pi(t) = \{\kappa_{ij}\}(t) \tag{19}$$

becomes time dependent; whether an element e_i does or doesn't belong to the system is expressed by a time variable relation with $\mu(e_i)(t)$. The fuzzy measure becomes time dependent. Regarding changes in environment like in a ecosystem, or observing a taking over (usurpation) of one system by another one, another way to express the change could be to alter the dimension of the state space and of the input and output vectors. The change of the characteristics of a behavior can be modeled by introducing control parameters, i.e., to make the system description of a chaotic type (pendulum of Poincaré).

Fig. 8 A ball is running on a stretched rubber sheet. The *squares* indicate a still flat surface in phase 1. In phase 2, the gravity of the ball is deforming the stretched surface, until the ball's velocity decreases and a crater is formed when its kinetic energy has been exhausted in friction and in energy forming the crater. The crater can be considered as a new attractor for the ball (cf. [66, p. 181])

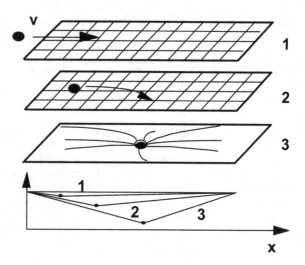

4.2 The Concept of Non-classical Systems

The concept of non-classicity originally comes from quantum mechanics and expresses that the state spaces are no simple spaces like in classical mechanics but dual spaces like Hilbert spaces, such that the sum of two state forms also a state (linear superposition). Moreover the interaction with an observer moves the system in a definable state whereas only the probability distribution for such possible states can be given by the theory. After the measurementprocess the system is in a defined state (due to the so-called collapse of wave vector), but this cannot be predicted exactly. The resulting lack of information, when describing the system in such way, is not removable and can be expressed by the famous uncertainty relation of Heisenberg. As a symptom we have by principle the appearance of complementarily related observables, say x and y, which cannot be measured with arbitrary exactness (e.g. x = location, y = impact or x = energy, y = time). The relevant operators don't commute, i.e.,

$$(O_x \bullet O_y - O_y \bullet O_x) \geq 0 \quad \text{with } \dim[O_x \bullet O_y] = \text{energy} \cdot \text{time} , \qquad (20)$$

i.e., the commutator has the physical dimension of action. (For more details see also [65].)

It has been discussed in quantum theory in the middle of 1980s in the last century whether the quantum mechanics is a axiomatic reduced system theory.[43] Taking

---

[43]I refer here to several oral communications in talks with F. H. Zucker in Cambridge, M.A. in 1981 and 1982. Cf. also some hints in [113] and presumably in [114]. I have not yet found an extensive exploration of this hypothesis within literature, but this could be an interesting research question.

the standpoint of descriptionism the meaning of this conjecture is that the question whether the quantum world per se would be a world by chance is no more relevant. Quantum mechanics as a system-theoretical non-classical description of the micro-world allows only a description on this level up to probabilities. Nevertheless the probability distribution can be predicted very exactly (spectra of possible values of observables and he probability distribution when measured frequently). We have to do it with lack of information in principle.

Further on there are field of objects in which a system-theoretical description forces us to introduce non-classical, i.e., complementarily related observables. Table 5 shows some examples from diverse fields of objects.[44]

Examples for a classical theory are the descriptions of classical mechanics, wave optics, electromagnetism (Maxwell's equations), general and special theory of relativity, theory of linear circuits, linear cybernetics like control theory without influencing observer. A classical theory can be characterized by the properties locality (L), determinacy (D), and predictability (P). In terms of descriptivism these are properties of the system description and not properties in the field of objects.

The principle of locality could also be named the principle of complete distinction, since one can always decide whether an object as a part of a system has a certain property or not. Further on it is always decidable whether the whole system is in a defined state or not. The states can always be separated from each other, and the system is only in one state at the same time.

A typical violation of L can be found in quantum theory and the attempt to remove it by the theory of hidden variables [10]. The non-locality in quantum theory leads to the fact that a localization of an actual state of the system in the state space is not possible anymore. The theory of hidden variables tries to avoid that by arguing that the collapse of the quantum mechanical wave vector into a defined, later measured state is determined completely by variables which are independent of time and space and which are not available for the system description in principle.

The principle of determinacy D assumes that each event has at least one cause (principle of sufficient causation), and each cause delivers at least one action. A violation of this principle we call chance. It can be interpreted in three ways:

1. There is a contingent, i.e., not necessary lack of information about all sufficient causes, like the Brownian motion of molecules, which could be removed in principle. If one states that

$$\frac{1}{2} m v^2 = \frac{3}{2} k T \qquad (21)$$

---

[44] A detailed discussion of such complementarily related observables has been given by Röhrle [99]. He tried to show the hypothesis that the appearance of complementarily related entities within a system description indicates sign that the field of interest is not yet satisfyingly understood.

Table 5 Complementarily
related observable entities in
several field of objects

	Observable x	Observable y
Field		
Pragmatic information novelty	Confirmation	Novelty
System theory	Structure	Behavior
Theory of reliability	Reliability	Autonomy
Theory of economy biology	Stability	Growth
Organization	Justification	Efficiency
Epistemology	Exo-view	Endo-view
Linguistics	Description	Interpretation

in thermodynamics one could measure for each molecule a velocity vi and calculate the mean

$$\frac{1}{n} \cdot \sum_{i}^{n} v_i = v. \tag{22}$$

Indeed, one is forced to withdraw on practical reasons. Chance is then a manner to characterize this gap within the description.

2. There is no removable lack of information, over all sufficient causes, since the description is incomplete in principle due to its complexity.[45]
3. Processes are possible without sufficient causes; nature per se is ruled by chance. This interpretation presupposes that it is possible to make statements about the ontological status of nature. This could be called a naturalization of chance. Within the realm of descriptionism this statement is not meaningful.

The principle of predictability P starts with the assumption that knowing all sufficient causes and the forgoing processes one should be able to predict at least one event. Now, predictions are only derivable from system descriptions if one has either a law of state transition, the initial state as well as the limitations of the state space (in physics these are the natural laws and constraints) or one makes an extrapolation from the already existing data of the system behavior, using knowledge about certain limitation of behavioral type (Kalman-Filtering, least square fit, exponential smoothing, etc.). The latter is possible without knowing laws. In the first case we would speak about a certain understanding of system dynamic, in the latter case we use only the trust into the limitation of the variance of a process.

The property of predictability P is violated by the lack of sufficient exact propositions about the initial and boundary conditions. Further on, if a process runs beyond the area of validity, this may lower the predictability, too. In case of extrapolation, the knowledge about an only short run decreases the quality of prediction. Accepting still an unlimited variation of the process (*natura facit saltae*),

---

[45]This has been shown by Kornwachs et al. [63, 71, 72] using a measure of complexity which is defined by the degree of time dependence of system structure.

any prediction will become impossible. This holds for the case as well, that the laws are incompletely known. In other words: chance and chaos are unpredictable by definition.

4.3   The LDP-Classification

After this short discussion it is possible to arrange the different degrees of violation of the properties L, P, and D and to get a "measure" for the violation of degree of being "classic."

The theories resp. system descriptions are listed in Table 6. The comments therein refer to the different forms of "non-classicity." In cases (1) and (2), a system surface could be defined according to the determining (stochastically or not) variables and independent variables. In cases (3) and (4) the borderline between quantum objects and observer (including the classical measurement device) can be shifted—this is named the Heisenberg Cut.[46] Synergetic or chaotic (i.e., nonlinear) systems (5) can be described with defined surfaces (as physical systems) like nonequilibrium thermodynamic systems (6), since the generation of new structures is due to the change of the decisive (control) parameters which can be regarded as constraints.

In case (7) Roger Penrose has introduced the distinction between weak and strong determinism [96, p. 422]. In the case of strong determinism it is postulated that the whole history of universe is fixed to all times. This can be expressed by the picture of Feymann-graphs of elementary processes in particle physics (physical states and time space). Even Albert Einstein was a proponent of this strong determinism; nevertheless he didn't see any contradiction to the free will of man. Also the Many World Interpretation of Quantum Mechanics may be counted to the class of strongly deterministic theory. The weak determinism is characterized by the assumption that the dynamics of the world is not fixed for all times but determined by preceding and present events. That is, an actual process determines the succeeding process iteratively, but each process has some uncertainties. The determination of the succeeding process is highly sensible with respect to his uncertainty. Thus the possibility to define a surface remains uncertain, too.

In case (8) we have the strongest violation of the LPD-properties. With respect to calculability it is possible to make a cross characteristic like in Table 7.

We may shortly discuss these four fields:

Machines

The paradigm of a machine is based on even this classical system description. The usual dynamics for continuous variables (or observables) is given by the equations of motion, as known in classical Hamilton dynamics. It has been proved

---

[46]Cf. [46]. For modern discussion see, e.g., [7, 13].

Table 6 An LDP—classification of non-classical system descriptions (resp. theories)

Case	Theory resp. system description	L	D	P	Comments
1	Classical systems linear system theory	+	+	+	Linear system theory
2	Correlations, QM-interpretation: Hidden variables	v	+	+	All non-local theories, e.g., statistical correlations, space- and time-independent variables determine the events
3	QM interpretation: Kopenhagen	v	+	v	The superposition prevents the locality of a theory; this leads to the probability interpretation of the collapse of the wave function[a] and the reduction of predictability.
5	QM many-World-interpretation, complex systems	+	v	+ up to $p(x)$	Prediction is only possible with probabilities $p(x)$; no deterministic process, since the description is not completely defined
6	Non-equilibrium-thermodynamics	+	v	v	Statistical "nature" of thermodynamic processes, neither deterministic nor predictable
7	R process (Penrose)[b] up to $p(x)$; symbolic processing in real (physical) systems	v	v	+ up to $p(x)$	Collapse of wave function in terms of Penrose—interpretation. Due to the irreversibility of data processing, operative efficiency of a program is not generally predictable[c].
8	Non-algorithmical processes in brain, social systems—living systems?	v	v	v	Non-local, nondeterministic and non-predictable theories; i.e., maximal violation of the property of being classic

(+) = principle is fulfilled; (v) = principle is violated
[a]When interacting with the observer
[b]Cf. [96]
[c]For problems of a semantic of computer programs see, e.g., [1] and the original papers by Floyd and Hoare [24, 48]

to be useful to distinguish between Newtonian mechanics whereas the particles have no inner interaction and the analytical mechanics according to Euler, Lagrange, and Laplace, which is determined by constraints. These constraints can be classified whether the respective differential forms can be integrated or not. If yes, the system shows holonome constraints. As an example, the definition of rigidity of a stone

Table 7 Classification of processes

	Processes	
	Deterministic	Nondeterministic
Algorithmic	Machines	(b)
Non-algorithmic	(a)	Ordinary chance

may serve, given by n particles with fixed distances. If the differential forms are not able to be integrated, we have non-holonome constraints.

In Newtonian mechanics we have a six-dimensional phase space, since the velocity

$$\frac{d\mathbf{x}}{dt} \tag{23}$$

of a particle is independent of the location vector—also expressable as a continuous state space—$\underline{\mathbf{x}}$, and the acceleration

$$\frac{d^2\underline{\mathbf{x}}}{dt^2} \tag{24}$$

depends only on a function of the phase (or state). The equation of motion is written as

$$\frac{d^2\underline{\mathbf{x}}}{dt^2} = F\left(\frac{d\underline{\mathbf{x}}}{dt}, t, \underline{\mathbf{x}}\right). \tag{25}$$

The function F expresses the respective forces, divided by a constant factor (in classical mechanics it is the mass m).

Having holonome constraints one can reduce the dimensionality of the state space to $n = 2$. With non-holonome constraints it is always locally possible to eliminate the velocity

$$\underline{\mathbf{v}} = \frac{d\underline{\mathbf{x}}}{dt} . \tag{26}$$

Partially constrained systems are called according to Haken [36] synergetic systems, whereas the maximal possible number of constraints leads to the component of the velocity vector

$$\frac{dx_i}{dt} = f_i(x_1, \ldots, x_n) \quad \text{with} \quad i = 1, \ldots, n \tag{27}$$

particles and n possible constraints. Rosen [101] has considered such maximal constrained systems as an explicit possibility to characterize a machine as a physical system.

Rosen [101] further showed that for a maximal constrained system with a given set of parameters, describing the coupling with external forces (like masses and others) as

$$\frac{dx_i}{dt} = f_i(x_1, \ldots, x_n, \beta_1, \ldots, \beta_n) \tag{28}$$

with a parameter $\beta \in B$ from the parameter space B, for a given state $x \in X$, each $\beta \in B$ defines a single tangential vector of a tangent space $T(X)$ at the point x. Therefore such parameters β can be conceptualized as control or input for a machine. The task of a control is to find an appropriate $x(t) \in T$ for a given $\beta(t)$. The main question is: what the system is doing when influenced by a given input or control. Another question is to find an appropriate control $\beta(t)$ such that $X(t)$ may have some defined or aimed optimal properties.

The decisive point by Rosen [101] is to use the locality of the system. The equation above, written in vector presentation

$$\underline{v}(t) = \frac{d\underline{x}}{dt} = f_i(\underline{x}, \underline{\beta}) \tag{29}$$

can be regarded as a local relation, i.e., it is only valid within the neighborhood of a state in which the system actually can be found. This state is such that for the respective components of the parameter vector (i.e., the control $\underline{\beta}$) holds

$$\beta_i(t) = \beta_i(\underline{x}, t) . \tag{30}$$

In this case, the control vector $\underline{\beta}$ may contain further parameters independent of time and the actual state. Let be $\overline{\gamma}$ be such a parameter vector, we can write

$$\beta_i(t) = \beta_i(\underline{x}, t, \gamma_j) . \tag{31}$$

Rosen [101] calls such a set of independent parameters $(\ldots, \gamma_j, \ldots)$ a program. A machine as a maximal or partial constrained system can be made to a programmable machine due to the impact of a control

$$\beta_i(t) = \beta_i(\underline{x}, t, \gamma_j) , \tag{32}$$

which is invariant with respect to time and state. This program could be interpreted as a surface of such a device when described as a system.[47]

---

[47]Nevertheless, it is not yet clear how the interaction between variables dependent upon time and location and invariant variables can be explained in an ontologically satisfying manner. From a phenomenological point of view one can model this within a system description if one has an idea how this interaction could be hypothesized. Cf. [64, 101].

Ordinary chance

In this case one is forced to apply statistical descriptions and to calibrate the probability distributions with experiences. In order to define probability at all one has to limit the number of possible states, items, or events; otherwise one is not able to normalize the probability distribution to

$$\sum_{i=1}^{n} p(x_i) = 1 .$$

(33)

This limitation can also be conceived as a definition of a surface.

Nevertheless, the interesting fields for a possible theory building could be found in the following items:

(a) Deterministic processes, which are not able to be modeled according to algorithms, e.g., evolution, growth, and decay of biological systems.[48]
(b) Nondeterministic processes, which one would like to model in terms of algorithm like free will, social dynamics, and cognition. It is required to find a suitable level of description, sufficiently aggregated in order to model again a deterministic state transition dynamics.[49]

It should be emphasized that the LDP classification doesn't say anything about the world but tells us something about ways how to describe systems, habitually used in science. It serves to localize interesting research topics in system theory and cybernetics. Moreover, it shows us how the deficit of the classical concept forces us to reduce the strong definition of surface toward a more flexible concept: the observer is the author of the surface.

Acknowledgments This paper is based on a presentation, held at Philosophisches Kolloquium, December 3rd, 2008 at the Institute for Philosophy, University of Jena, Frege Centre for Structural Sciences. The inspiring discussions with Bernd-Olaf Küppers and Stefan Artmann are gratefully acknowledged.

---

[48]It is really astonishing that processes with changing structures and patterns of behavior like decay and growth cannot be modeled satisfyingly with usual system theoretical descriptions, based on elementary set theory and mappings. This would require designing sets mathematically, which can generate or annihilate their own elements without a choice process from another finite set, of already "stored" elements. There is a correspondence to this problem with the fact that there is no real calculable GOTO in the theory of algorithms.

[49]To introduce a wave function ψ in quantum mechanics is an actual strategy to do so. As a state ψ is no observable, but it obeys a linear "equation of motion," the well-known Schrödinger equation. Even mathematical operations on this wave function will allow a calculation of the possible values (*eigenvalues*) of observables.

References

1. Allison, L.: A Practical Introduction to Denotational Semantics. Cambridge University Press, Cambridge (1986)
2. Apel, K.O.: Transformation der Philosophie, 2 Bde, 2nd edn. Suhrkamp, Frankfurt a.M. (1976)
3. Aristoteles: Vom Himmel. Von der Seele. Von der Dichtkunst. Übertragen und eingeleitet von O. Gigon, Hrsg. v. H. Hoenn. Artemis, Zürich (1950)
4. Aristotele: De Generatione et Corruptione. Über Werden und Vergehen. Hrsg. von Th. Buchheim. Meiner, Hamburg (2011)
5. Aristoteles: Nikomachische Ethik. (Übersetzt von O. Gigon). Artemis, Zürich (1967)
6. Atmanspacher, H.: Objectivation as an endo-exo transition. In: Atmanspacher, H., Dalenoort, G.J. (eds.) Inside Versus Outside. Endo- and Exo-Concepts of Observation and Knowledge in Physics, Philosophy and Cognitive Science, pp. 15–32. Springer, Berlin (1994)
7. Atmanspacher, H.: Cartesian cut, Heisenberg cut, and the concept of complexity. World Futures **49**, 333–355 (1997)
8. Bellmann, R.E.: Dynamische Programmierung und selbstanpassende Regelprozesse. Oldenbourg, München (1967)
9. von Bertalanffy, L.: General System Theory. Penguin Books, Hammondsworth (1968/1973)
10. Bohm, D., Bub, J.: A proposed solution of the measurement problem in quantum mechanics by a hidden variable theory. Rev. Mod. Phys. **38**, 453 ff (1966)
11. Bohrmann, A.: Bahnen künstlicher Satelliten. BI, Mannheim (1966)
12. Bonabeau, E., Dorigo, M., Theraulaz, G.: Swarm Intelligence: From Natural to Artificial Systems (Santa Fe Institute Studies in the Sciences of Complexity Proceedings) (Paperback). Oxford University Press, London (1999)
13. von Borzeszkowski, H.-H., Wahsner, R.: Heisenberg's cut and the limitations of its movability. Annalen der Physik **500**, 522–528 (1988)
14. Bullinger, H.-J., Kornwachs, K.: Expertensysteme – Anwendungen und Auswirkungen im Produktionsbetrieb. C.H. Beck, München (1990)
15. Capelle, W.: Die Vorsokratiker. Kröner, Stuttgart (1963)
16. Carnap, R.: Einführung in die Philosophie der Naturwissenschaft. Nymphenburger, München (1969)
17. Carnap, R.: Scheinprobleme in der Philosophie. Suhrkamp, Frankfurt/Main (1971)
18. Chaitin, G.J.: Information theoretic computational complexity. IEEE Trans. Inform. Theor. **20**, 10–15 (1974)
19. Chaitin, G.J.: Gödel's theorem and information. Int. J. Theoret. Phys. **21**(12), 941–954 (1982)
20. Chomsky, N.: Aspects of the Theory of Syntax. MIT Press, Cambridge (1965)
21. Descartes, R.: Die Prinzipien der Philosophie (1644). Meiner, Hamburg (1965)
22. Diels, H., Kranz, W. (eds.): Die Fragmente der Vorsokratiker, 6th edn. B Fragmente, Dublin, Zürich (1972)
23. Einstein, A.: Geometrie und Erfahrung (1921). In: Einstein, A. (ed.) Mein Weltbild, pp. 119–127. Ullstein, Frankfurt a.M. (1983)
24. Floyd, R.W.: Assigning meanings to programs. In: Schwartz, J.T. (ed.) Mathematical Aspects of Cognitive Science. Proceedings of Symposium on Applied Mathematics, vol. 19, pp. 19–32. AMS, New York (1967)
25. von Foerster, H.: On constructing a reality. In: Preiser, W.F.E. (ed.): Environmental Design Research, vol. 2, pp. 35–46. Dowden, Hutchinson & Ross, Stroudsburg (1973)
26. von Foerster, H.: Objects: Tokens for (Eigen-)Values. Cybern. Forum **8**, 91 (1976)
27. Frege, G.: Logik in der Mathematik. In: Nachgelassene Schriften, hrsg. von H. Hermes, F. Kambartel und F. Kaulbach, 1. Band. Meiner, Hamburg (1969)
28. Gallee, M.A.: Bausteine einer abduktiven Wissenschafts- und Technikphilosophie. Lit, Münster (2003)

29. Gent, H.: Summa Quaestionum Ordinarium, vol. 2 (reprint). Franciscan Institute Press, Bonaventura (1953)
30. Glansdorff, P., Prigogine, I.: Structure, Stabilité et Fluctuations. Masson, Paris. English: Thermodynamic Theory of Structure. Stability and Fluctuations. Wiley, London (1971)
31. Gödel, K.: Formal unentscheidbare Sätze in der principia mathematica. Monatshefte für Mathematik und Physik **38**, 173–198 (1931)
32. Grillparzer, F.: Epigramme. Hanser, München (1960)
33. Habermas, J.: Erkenntnis und Interesse. Suhrkamp, Frankfurt/Main (1971)
34. Habermas, J.: Theorie der Gesellschaft oder Sozialtechnologie? Einleitung. In: Habermas, J., Luhmann, N. (eds.) Theorie der Gesellschaft oder Sozialtechnologie, pp. 142 ff. Suhrkamp, Frankfurt/Main (1971)
35. Hager, F.-P.: System. I. Antike. In: Ritter, J., Gründer, K. (eds.) Historisches Wörterbuch der Philosophie, Bd. 10, pp. 824–825. Wiss. Buchgesellschaft, Darmstadt (1998)
36. Haken, H.: Synergetics. An Introduction. Springer, Berlin (1978)
37. Haken, H.: Synergetics and the problem of Selforganization. In: Roth, G., Schwegler, R. (eds.) Selforganizing Systems. Campus Verlag, Frankfurt (1979)
38. Hegel, G.W.F.: Phänomenologie des Geistes. Vorrede (Bamberg, Würzburg, 1807), pp. 11–15. Ullstein, Frankfurt/Main (1970)
39. Hegel, G.W.F.: Grundlinien der Philosophie des Rechts. Vorrede (Berlin, 1821). In: Hauptwerke in sechs Bänden, Bd. 5. Meiner, Hamburg (1999)
40. Heidegger, M.: Identität und Differenz. Neske, Pfullingen (1957)
41. Heidegger, M.: Sein und Zeit, 11. Aufl. Niemeyer, Tübingen (1967). Engl.: Being and Time; translated by J.M. Macquarrie and E. Robinson. Blackwell, New York (1962), Reprint 2005
42. Heidegger, M.: Nur ein Gott kann uns retten. Der Spiegel **23**, 209 (1976)
43. Heisenberg, W.: Über quantentheoretische Umdeutung kinematischer und mechanischer Beziehungen. Zeitschrift für Physik **33**, 879–893 (1925)
44. Heisenberg, W.: Physik und Philosophie. Hirzel, Stuttgart (1962); Engl.: Physics and Philosophy: The Revolution in Modern Science (1958). Lectures delivered at University of St. Andrews, Scotland, Winter 1955–1956. Harper & Broth, New York (1958), Reprint 2007
45. Heisenberg, W.: Der Teil und das Ganze. Piper, München (1969)
46. Heisenberg, W.: Ist eine deterministische Ergänzung der Quantenmechanik möglich? (1935). In: Von Meyenn, K. (ed.) Wolfgang Pauli: Scientific Correspondence with Bohr, Einstein, Heisenberg, vol. II, pp. 409–418. Springer, Berlin (1985)
47. Helbing, D., Farkas, I., Vicsek, T.: Simulating dynamical features of escape panic. Nature **407**, 487–490 (2000)
48. Hoare, C.A.R.: An axiomatic basis for computer programming. Comm. ACM **12**(10), 576–580, 583 (1969)
49. Hubig, C.: Mittel. Transcript, Bielfeld (2002)
50. Hübner, W.: Ordnung. II. Mittelalter. In: Ritter, J., Gründer, K. (eds.) Historisches Wörterbuch der Philosophie, Bd. 6, Sp. 1254–1279. Wiss. Buchgesellschaft, Darmstadt (1984)
51. Kalmann, R.E.: On the general theory of control systems. In: Proceedings of IFAC Congress, Moskau, Oldenburg, München (1960)
52. Kant, I.: Allgemeine Naturgeschichte und Theorie des Himmels oder Versuch von der Verfassung und dem mechanischen Ursprung des ganzen Weltgebäudes, nach Newtonschen Grundsätzen abgehandelt (Königsberg, Leipzig, 1755). In: Gesammelte Schriften, hrsg. von der Kgl. Preu. Akad. der Wissenschaften. Georg Reimer, Berlin (1910)
53. Kant, I.: Kritik der reinen Vernunft. Riga 1781, Bd. 37a. Felix Meiner, Hamburg (1956)
54. Kant, I.: Kritik der Reinen Vernunft. Riga 1978. In: Werke. (Akademieausgabe), hrsg. von Wilhelm Weischedel, Norbert Hinske. 6 Bde. Suhrkamp, Frankfurt am Main (1968)
55. Klir, G.J.: An Approach to General System Theory, p. 282. Van Nostrand, New York (1969)
56. Klix, F.: Information und Verhalten. VEB Deutscher Verlag der Wissenschaften, Berlin (1973)
57. Kolmogorov, A.N.: Foundation of Probability. Chelsea Publishing, New York (1950)
58. Kolmogorov, A.N.: Three approaches to the quantitative definition of information. Prob. Inform. Transm. **1**, 1–7 (1965)

59. Kornwachs, K.: Technik impliziert ihre Verwendung. Bild der Wissenschaft, Heft **9**, 120–121 (1980)
60. Kornwachs, K.: Function and information - towards a system-theoretical description of the use of products. Angewandte Systemanalyse **5**(2), 73–83 (1984)
61. Kornwachs, K. (Hrsg.): Offenheit - Zeitlichkeit - Komplexität. Zur Theorie der offenen Systeme. Campus, Frankfurt (1984)
62. Kornwachs, K.: Utility and costs of information. A systematical approach to an information-cost-estimation. Angewandte Systemanalyse **5**(3–4), 112–121 (1984)
63. Kornwachs, K.: Offene Systeme und die Frage nach der Information. Habilitation Thesis, University of Stuttgart (1987)
64. Kornwachs, K.: Complementarity and cognition. In: Carvallo, M.E. (ed.) Nature, Cognition and Systems, pp. 95–127. Kluwer, Amsterdam (1988)
65. Kornwachs, K.: Information und der Begriff der Wirkung. In: Krönig, D., Lang, R. (Hrsg.) Physik und Informatik – Informatik und Physik. Informatik Fachberichte, Nr. 306, S. 46–56. Springer, Berlin (1991)
66. Kornwachs, K.: Pragmatic information and system surface. In: Kornwachs, K., Jacoby, K. (eds.) Information – New Questions to a Multidisciplinary Concept, pp. 163–185. Akademie, Berlin (1996)
67. Kornwachs, K.: Pragmatic information and the emergence of meaning. In: Van de Vijver, G., Salthe, S., Delpos, M. (eds.) Evolutionary Systems, pp. 181–196. Boston, Kluwer (1998)
68. Kornwachs, K.: System ontology and descriptionism: Bertalanffy's view and new developments. TripleC **2**(1), 47–62 (2004)
69. Kornwachs, K.: Vulnerability of converging technologies – the example of ubiquitous computing. In: Banse, G., Hronszky, I. (eds.) Converging Technologies, pp. 55–88. Edition Sigma, Berlin (2007)
70. Kornwachs, K.: Nichtklassische Systeme und das Problem der Emergenz. In: Breuninger, R., et al. (Hrsg.) Selbstorganisation. 7. Ulmer Humboldt-Kolloquium. Bausteine der Philosophie Bd. 28, S. 181–231. Universitätsverlag, Ulm (2008)
71. Kornwachs, K., von Lucadou, W.: Komplexe Systeme. In: Kornwachs, K. (Hrsg.) Offenheit – Zeitlichkeit – Komplexität, pp. 110–165. Campus, Frankfurt (1984)
72. Kornwachs, K., von Lucadou, W.: Open systems and complexity. In: Dalenoort, G. (ed.) The Paradigm of Self Organization, pp. 123–145. Gordon & Breach, London (1989)
73. Kuhn, T.S.: Die Struktur wissenschaftlicher Revolution, 2nd edn. Suhrkamp, Frankfurt/Main (1973)
74. Küppers, B.-O.: Die Strukturwissenschaften als Bindeglied zwischen Natur- und Geisteswissenschaften. In: Küppers, B.-O. (Hrsg.) Die Einheit der Wirklichkeit – Zum Wissenschaftsverständnis der Gegenwart, S. 89–106. W. Fink, München (2000)
75. Kuznetsov, Yu.A.: Elements of Applied Bifurcation Theory. Springer, New York (1995)
76. Landau, L.D., Liefschitz, E.M.: Lehrbuch der Theoretischen Physik. Bd. III: Quantenmechanik. Akademie verlag, Berlin (Ost) (1967)
77. Lazlo, E., Margenau, H.: The emergence of integrative concepts in contemporary science. Philos. Sci. **39**(2), 252–259 (1972)
78. Leibniz, G.W.: Monadologie. In: Leibniz, G.W. (ed.) Hauptwerke. Hrsg. von G. Krüger, pp. 131–150. Kröner Verlag, Stuttgart (1958)
79. Lenk, H.: Wissenschaftstheorie und Systemtheorie. In: Lenk, H., Ropohl, G. (eds.) Systemtheorie als Wissenschaftsprogramm, S. 239–269. Athenäum-Verl., Königstein (1978)
80. Locke, M.: Grundlagen einer Theorie allgemeiner dynamischer Systeme. Akademie verlag, Berlin (Ost) (1984)
81. Lorenz, E.N.: Deterministic nonperiodic flow. J. Atmos. Sci. **20**(2), 130–141 (1963)
82. von Lucadou, W., Kornwachs, K.: The problem of reductionism from a system theoretical viewpoint - how to link Physics and Psychology. Zeitschrift für Allgemeine Wissenschaftstheorie **XIV/2**, 338–349 (1983)
83. Luhmann, N.: Soziale Systeme. Suhrkamp, Frankfurt/Main (1984)

84. McCulloch, W.S., Pitts, W.: A logical calculus of the ideas immanent in nervous activities. Bull. Math. Biophys. **5**, 115–133 (1943)
85. Meixner, J.: Zur Thermodynamik der irreversiblen Prozesse. Zeitschrift für physikalische Chemie (B) **53**, 235, 238–242 (1943)
86. Mesarovic, M.D.: A mathematical theory of general systems. In: Klir, G. (ed.) Trends in General System Theory, pp. 251–269. Wiley-Interscience, New York (1972)
87. Morin, K., Michalski, K., Rudolph, E. (eds.): Offene Systeme II – Logik und Zeit. Klett-Cotta, Stuttgart (1981)
88. von Neumann, J.: The general and logical theory of automata. In: Jeffers, L.A. (ed.) Cerebral Mechanism of the Behavior. Wiley, New York (1951). Dt.: Allgemeine und logische Theorie der Automaten. Übersetzt von D. Krönig. In: Kursbuch 8, März, pp. 139–192 (1967)
89. von Neumann, J.: The Computer and the Brain (8th printing). Yale University Press, New Haven (1974)
90. von Neumann, J., Morgenstern, O.: Spieltheorie und wissenschaftliches Verhalten. Physica, Würzburg (1961)
91. Neumann, J.V.: In: Burks, A. (ed.) Theory of Self-Reproducing Automata. Urbana, London (1966)
92. Newell, A., Simon, H.A.: Human Problem Solving. Prentice Hall, Englewood Cliffs (1972)
93. Padulo, L., Arbib, M.A.: System Theory - A Unified State Space Approach to Continuous and Discrete Systems. Hemisphere Publ., Washington (1974)
94. Palm, G.: Neural Assemblies. Springer, Berlin (1981)
95. Parson, T.: The Social System. Free Press, New York (1957)
96. Penrose, R.: Computerdenken- die Debatte um Künstliche Intelligenz, Bewusstsein und die Gesetz der Physik. Spektrum, Heidelberg (1991)
97. Pontrjagin, L.S., Boltjanskij, V.G., Gamkrelidze, R.V., Miscenko, E.F.: Mathematische Theorie optionaler Prozesse. VEB Deutscher Verlag der Wissenschaften, Berlin (Ost) (1964)
98. Prigogine, J.: Vom Sein zum Werden. Piper, München (1979)
99. Röhrle, E.A.: Komplementarität und Erkenntnis. – von der Physik zur Philosophie. Lit, Münster (2000)
100. Ropohl, G.: Systemtheorie der Technik. Hanser, München (1979); 2nd edition: Allgemeine Technologie. Hanser, München (1999); 3rd edition: Karlsuher Universitätsverlag, Karlsruhe (2009)
101. Rosen, R.: Causal structures in brains and machines. Int. J. Gen. Syst. **12**, 107–116 (1986)
102. Schulz, R.: Systeme, biologische. In: Ritter, J., Gründer, K. (eds.) Historisches Wörterbuch der Philosophie, Bd. 10, pp. 856–862. Wiss. Buchgesellschaft, Darmstadt (1998)
103. Shannon, C.E., Weaver, W.: The Mathematical Theory of Communication. Urbana, Chicago, London (1949/1969). Deutsch: Mathematische Grundlagen der Informationstheorie. R. Oldenbourg, München (1976)
104. Strub, Ch.: System. II. System und Systemkritik in der Neuzeit. In: Ritter, J., Gründer, K. (eds.) (1984) Historisches Wörterbuch der Philosophie, Bd. 10, pp. 25–856. Wiss. Buchgesellschaft, Darmstadt (1998)
105. Thom, R.: Structural Stability and Morphogenetics. Benjamin, Reading (1975)
106. Weizenbaum, J.: Die Macht des Computers und die Ohnmacht der Vernunft. Suhrkamp, Frankfurt a.M. (1977)
107. von Weizsäcker, C.F.: Die Einheit der Natur, S. 352. Hanser Verlag, München (1971)
108. von Weizsäcker, E.U.: Erstmaligkeit und Bestätigung als Komponenten der Pragmatischen Information. In: von Weizsäcker, E.U. (Hrsg.) Offene Systeme I, S. 82–113. Klett, Stuttgart (1974)
109. Wiener, N.: Kybernetik – Regelung und Nachrichtenübertragung in Lebewesen und Maschinen. Rowohlt, Reinbeck, 1968 (1971), Econ Düsseldorf, Wien (1963), neue Auflage (1980)
110. Winston, P.H.: Artificial Intelligence, 3rd edn. Addison-Wesley, Reading (1992)
111. Wunsch, G.: Geschichte der Systemtheorie – Dynamische Systeme und Prozesse, Oldenbourg, München, Wien (1985)

112. Zahn, M.: System. In: Krings, H., Baumgartner, H.M., Wild, Ch. (eds.) Handbuch
     Philosophischer Grundbegriffe, Band 5, S. 1458–1475. Kösel Verlag, München (1974)
113. Zucker, H.F.: Information, Entropie, Komplementarität und Zeit. In: von Weizsäcker, E.U.
     (eds.) Offene Systeme I, S. 35–81. Klett, Stuttgart (1974)
114. Zucker, H.F.: Phenomenological Evidence and the "Ideal" of Physics. Working Paper, Max
     Planck Institut zur Erforschung der Lebensbedingungen, Starnberg, Probable (1978)

Elements of a Semantic Code

Bernd-Olaf Küppers

Abstract An important step towards a scientific understanding of complex systems will be the objectification and quantification of information that carries semantics, i.e., sense and meaning. In this paper a concept is developed according to which the general aspects of semantics—such as novelty, pragmatic relevance, selectivity, complexity and others—are understood as elements of a "semantic code". However, in contrast to its traditional usage, the term "code" does not refer to a set of rules for assignment or translation of symbols, but rather to a reservoir of value elements, from which the recipient configures a value scale for the evaluation of semantic information. A quantitative measure for the value scale is proposed.

1 Three Dimensions of Information

Information is based upon signs and symbols and sequences of these. Such sequences may be rich in content, that is, they can have a sense or a meaning.[1] The content, in turn, may become operative in various ways, because of the different kinds of reactions that it can induce in the recipient. In general, three dimensions of information can be distinguished. The "syntactic" dimension is understood as the ordered arrangement of symbols and the relationships between them. The "semantic" dimension includes the relationships between the symbols and also that for which they stand. Finally, the "pragmatic" dimension includes the relationships

---

[1] Here the two terms *sense* and *meaning* are used synonymously, and the distinction between them that is usual in linguistic philosophy is not retained.

B.-O. Küppers
Frege Centre for Structural Sciences, Friedrich Schiller University Jena, Jena, Germany
e-mail: bernd.kueppers@uni-jena.de

B.-O. Küppers et al. (eds.), *Evolution of Semantic Systems*,
DOI 10.1007/978-3-642-34997-3_4, © Springer-Verlag Berlin Heidelberg 2013

between the symbols, that for which they stand, and also the effect that they have upon the recipient.

However, resolving the concept of information into a syntactic, semantic and pragmatic dimension is only justified within the framework of a systematic analysis. Strictly speaking, the three dimensions make up an indissoluble unity, as one cannot allude to one of them on its own without automatically invoking the other two.

Syntax only takes on a meaning when both the symbols as such and their interrelationships have been stipulated. Semantics, in turn, presuppose a syntactic structure as carrier of the semantics, even though semantics go further than this. The same applies to pragmatics, since the sense and the meaning of information do not open up without reference of the symbols to the real world. Sense and meaning of information can only be objectified through the pragmatic effects that they have upon the recipient and which trigger his interactions with the surrounding world.

Because of the intimate intermeshing of semantics and pragmatics, they are frequently considered together as the so-called "semantico-pragmatic" aspect of information. However, the pragmatic aspect is accompanied by further facets of semantic information, expressed in its novelty, complexity, selectivity and other features. Nevertheless, it is purely a matter of definition whether one considers these as independent properties of semantic information or perhaps as various aspects of the pragmatic component that endow the latter with a fine structure. In any case the semantic aspect of information is something like a "hinge" between its syntactic and its pragmatic level.

The syntactic and the pragmatic aspects are ultimately equivalent to the structural and functional aspects of information. Even though function necessarily presupposes a structure as its carrier, this does not mean that functions can be traced back to their structures. Human language may provide a good example of this: here, the meaning and thus the pragmatic—i.e., functional—relevance of words and sentences are fixed through their syntax, but nonetheless cannot be derived from it. Rather, the semantics appear to stand in a "contingent" relationship to the syntax. In other words, the semantics of a syntactic structure are not necessarily the way they are. They are governed neither by chance nor by law.

Moreover, in human communication the exchange of semantic information is always an expression of human mental activities with all their individual and unique features. Can these be blanked out, allowing us to approach the general characteristics of semantic information? Above all: is there a semantic code that embodies the principles according to which meaningful information is constituted?

Questions of this kind appear to go well, and irrevocably, beyond the scope of the exact sciences, which are based upon objectification, idealisation, abstraction and generalisation. It is therefore understandable that the founding fathers of communication theory—Hartley, Shannon and others—were seeking a scientific approach to the concept of information that is independent of the *meaning* of information but that provides a measure of the *amount* of information.

In fact, communication technology is not concerned with the actual content of information; rather, it sees its primary task in passing the information in a reliable way from the sender to the receiver. The path of this passage, the so-called information channel, has to function such as to be nearly error-free as possible and to avoid corrupting the sequence of symbols on their way to the recipient. This presents the engineer with a technical challenge that is completely separate from the sense and the meaning the transmitted information may have for the recipient.

While it is true that communications engineers definitely speak of the "content" of information, they are referring not to its meaning, but rather to the sheer number of symbols that have to be transmitted. Thus, Shannon and Weaver state explicitly that: "In fact, two messages, one of which is heavily loaded with meaning and the other of which is pure nonsense, can be exactly equivalent (...) as regards information" [19].

According to our intuitive understanding, information has the task of eliminating uncertainty. This thought also underlies Shannon's definition of information [17]. According to it, the probability p_k with which the arrival of a particular message x_k is to be expected out of a set of possible messages may be considered as a simple measure of its information content. On the basis of this idea the information content of a message x_k is arranged inversely proportional to its expectation value p_k, or

$$I_k \sim 1/p_k. \tag{1}$$

If one demands from the information metric that, as a measure of a quantity, it should be additive, then one has to replace the above definition with

$$I_k = \mathrm{ld}(1/p_k) = -\mathrm{ld}\, p_k, \tag{2}$$

where we have introduced the binary (or dyadic) logarithm ld. This definition thus also takes account of the conventional use of binary coding in communications technology. If, in the simple case, a source of messages consists of N messages, which all have the same expectation value for the recipient, then the recipient needs

$$I = \mathrm{ld}(N) \tag{3}$$

binary decisions to select a given message. For this reason the quantity I is also termed the "decision content" of the message in question. In fact, the decision content had been proposed by Hartley [7] as a measure of the quantity of information some years before Shannon formulated the general expression above. Even though definition (2) does not consider the meaning of information, it points at least indirectly to a semantic aspect, namely that of the novelty of an item of information, insofar it is correlated with its inverse expectation probability.

The prior probabilities p_k are fixed by the recipient's prior knowledge. They are, however, objectively determined quantities, i.e., they are the same for all recipients that share the same facilities for, and methods of, obtaining knowledge. One can also say that the information measure developed by Shannon is "objectively subject-related" [20]. In other words: information is not an absolute entity, but only a relative one—a fact which is also termed the "context-dependence" of information.

Shannon's measure of information refers to a message source X of which the elements $\{x_1, \ldots, x_N\}$ make up a probability distribution $P = \{p_1, \ldots, p_N\}$. A characteristic feature of X is its statistical weight

$$H = \sum_k p_k I_k = -\sum_k p_k \, \mathrm{ld} \, p_k, \tag{4}$$

which corresponds to the mean information content of a message source. H is also termed the "entropy" of the source, because it has the same mathematical structure as the entropy function of statistical physics [see below, (11)].

An increase in information always takes place when through the arrival of a message the original distribution of expectation probabilities

$$P = \{p_1, \ldots, p_N\} \quad \text{with} \quad \sum_k p_k = 1 \quad \text{and} \quad p_k > 0 \tag{5}$$

becomes modified and converted into a new distribution

$$Q = \{q_1, \ldots, q_N\} \quad \text{with} \quad \sum_k q_k = 1 \quad \text{and} \quad q_k \geq 0. \tag{6}$$

If the gain of information is represented by the difference

$$H(P) - H(Q) = \sum_k p_k I_k - \sum_k q_k I_k \tag{7}$$

then this may lead to negative values, which would contradict our everyday picture of a gain, which we associate with a real increase in information. However, as Rényi [14] has shown, this deficiency can be resolved when one calculates the gain not by considering the difference between two mean values, but rather by first determining the gain in information for each single message and then calculating the mean value

$$H(Q|P) = \sum_k q_k [I_k(p_k) - I_k(q_k)] = \sum_k q_k \, \mathrm{ld}(q_k/p_k). \tag{8}$$

The Rényi entropy always fulfils the relation

$$H(Q|P) \geq 0. \tag{9}$$

Classical information theory, as developed on the basis of Hartley's and Shannon's work, always refers to a source of messages, which in turn is characterised by a probability distribution. Thus, the question automatically follows whether the information content of a given message can also be quantified without reference to its source. The basis for the measure of information would then be not a probability distribution, but rather a property of the message itself.

A measure of this kind has been proposed independently by Kolmogorov [8], Solomonoff [18] and Chaitin [4]. This approach has become known as algorithmic theory of complexity, since in it the concept of information is intimately linked to that of complexity. According to this idea, a sequence of binary digits is to be considered complex when the sequence cannot be compressed significantly, i.e., when there is no algorithm that is shorter than the sequence itself and from which the sequence can be derived. To express it in the terminology of computer science: the complexity K of a binary sequence S is given by the length L of the shortest program p of a computer C from which S can be generated:

$$K_C(S) = \min_{C(p)=S} L(p). \tag{10}$$

Thus, within this concept, the information content is defined through the complexity of a sequence of symbols. It provides, as it were, a measure of the "bulk" of the information contained in a message.

Like Shannon's definition of information, algorithmic information takes no account of the actual content of information. Rather, the algorithmic measure depends solely upon the complexity of an item of information, which in this case means the aperiodicity of its syntax. From a technological point of view, those restrictions are perfectly justified. However, in the natural sciences such a concept of information soon reaches the limits of its usefulness. This applies above all in the study of biological systems. This is because the results of modern biology, especially in the area of genomic analysis, show clearly that a profound understanding of biological phenomena requires access to the aspect of meaningful information. This applies not least to the problem of the origin of life itself, which is equivalent to that of the origin and evolution of information that has a meaning for the inception and sustenance of vital functions [9].

Insofar as syntax, semantics and pragmatics are intimately interrelated (albeit in ascending order of complexity), we are confronted in a fundamental way with the question of whether the semantico-pragmatic level of information is accessible to an exact, scientific analysis. Here—as we have already pointed out—we encounter major difficulties, because according to our traditional understanding of science the level of sense and meaning evades the kind of objectification and quantification that characterises the exact sciences.

The nature of these issues can be illustrated by the human language. As syntax, semantics and pragmatics represent increasingly complex dimensions of information, one's first thought might be to try and derive the semantics of a linguistic expression from the lower level of its syntax, i.e., from the sequence of phonemes, letters or other symbols which constitute this expression. However, this thought is short-lived, since there is no necessary connection between the syntax and the semantics of a message that would allow us to move logically from the one to the other. A garbled or incomplete text, for example, cannot in general be reconstructed by analysing the available fragments. This can at best be done against a background of additional information, which can serve as a semantic reference frame for the reconstruction.

In linguistics, the relationship between syntax and semantics is referred to as "arbitrary", because it does not allow any necessary connection to be discerned. According to this view, syntax is indeed a carrier of semantics, but semantics are not caused by the syntax. Rather, the irreducibility of semantics seems to be an example of "supervenience", i.e., the relationship of non-causal dependence between two properties A and B.

It seems to be a plausible explanation that the arbitrariness of the relationship between syntax and semantics is a consequence of a convention that is determined by the community using the language in question. Nonetheless, the use of the terms "arbitrary" and "conventional" can be criticised, as they still imply a causation principle, even if this is only in the form of the fixation of the semantics by the decision of the language community. For this reason, it might be better to describe the relationship between syntax and semantics as "contingent", as this term is neutral in respect of causation; it merely expresses the fact that the relationship between syntax and semantics is not of necessity that which one observes it to be.

Syntax and semantics are thus independent dimensions of information insofar as their interrelationship is not law-like, even though they are indissolubly connected to one another. Furthermore, the term "semantics" is strongly associated with numerous connotations such as sense, meaning, usefulness, functionality, value, content and so forth. These in turn leave a relatively large scope for determining the semantic dimension of information.

There are two possible ways of coping with this problem. One is to regard precisely the multiplicity as a characteristic feature of semantics. In that case, one would retain the variety of possible definitions and try to build up a theory of semantics on the most general basis possible. The other is to attempt to narrow down the broad field of possible definitions, in order to sharpen the concept of semantics. In that case one would arrive at a definition that would be precise, but at the cost of considerable one-sidedness.

In the latter respect, the greatest progress has been made by the analytical philosophy of language. This already possesses an elaborate theory of semantics, at the centre of which is the logical construction of human language. For example, within this theory strict distinction is made between concepts such as "sense" and "meaning", which in non-scientific usage are largely synonymous.

According to Frege, the founder of logical semantics, the "meaning" of a symbol or an expression is the object that it denotes, while the "sense" is the manner in which it is presented. Thus, the terms "evening star" and "morning star" have the same meaning, as they refer to the same object, the planet Venus; however, as Frege explains, they have different senses, because Venus is presented in the expression "morning star" differently from the way in which it is presented in the expression "evening star". The fact that the morning star and the evening star are the same object requires no further justification. However, the statement that the "star" we see close to the rising sun is the same "star" that appears in the evening sky after sunset is not an obvious truth. Rather, it is a discovery that we owe to the Babylonian astronomers. Semantic precision of this kind was applied by Frege not only to proper nouns, but also to entire sentences. Ultimately, they led to profound insights into the truth value of linguistic statements.

2 The Language of Genes

Even though logical depth is a unique property of human language, we may consider the possibility that language is a general principle of natural organisation that exists independently of human beings. In fact, there is much evidence in support of this hypothesis. Thus, it is generally accepted that various forms of communication exist already in the animal kingdom, as revealed by comparative behavioural research. We also know that even plants use a sophisticated communication system, employing certain scent molecules, to inform each other and in this way to protect against the threat of danger from pests. And it has long been known that bacteria communicate with one another by releasing chemical signals.

However, more surprising is the fact that essential characteristics of human language are reflected even in the structure of the genetic information-carriers. This analogy does not just consist in a mere vague correspondence; rather, it embraces largely identical features that are shared by all living beings.

Let us consider some facts: the carriers of genetic information, the nucleic acids, are built up from four classes of nucleotide, which are arranged in the molecules like the symbols of a language. Moreover, genetic information is organised hierarchically: each group of three nucleotides forms a code-word, which can be compared to a word in human language. The code-words are joined up into functional units, the genes. These correspond to sentences in human language. They in turn are linked up into chromosomes, which are higher-order functional units, comparable to a long text passage. As in a written language, the "genetic text" includes punctuation marks, which label—for example—the beginning and end of a unit to be read. Last but not least, the chemical structure of the nucleic acids even imposes a uniform reading direction.

In addition to the parallels described here, there are further fundamental correspondences between the structure and function of genes and that of human language. These include, especially, the vast aperiodicity of nucleotide sequences and the context-dependence of genetic information. The context in turn is provided by the physical and chemical environment, which confers an unambiguous sense upon the (in itself) plurivalent genetic information [10].

Just as a printing error in a written text can distort the meaning, the replacement of a single nucleotide in the genome can lead to collapse of the functional order and thus to the death and decay of the organism. This shows that genetic information also has a semantic dimension; in other words, it possesses functional significance for the sustenance of life processes. As the dynamics of life are encoded in the genes, it would seem only consistent to speak of the existence of a molecular language.

Naturally, the analogy between the language of genes and that of humans would immediately break down if we were to make the entire richness of human language the measure of our comparison with the genetic language. The language of Nature is more to be understood as an all-pervading natural phenomenon, which has found its most elementary form of expression in the language of the genes and its highest in human language.

The use of terms such as "genetic information" and "the language of genes" is in no way an illegitimate transfer of linguistic concepts to the non-linguistic realm of molecules. On the contrary: the existence of a genetic, molecular language appears to be an indispensable prerequisite for the construction of living systems, as their complex functional organisation arises along the path of material instruction and communication. In fact, the unfolding and expression of the tremendous amount of information stored in the genome takes place stepwise in the form of an exceedingly intricate process of communication between gene and gene product. However, any communication requires certain rules and these can only be understood by an appropriate reference to the model of language.

In view of the broad-ranging parallels between the structures of human language and the language of genes, recent years have seen even the linguistic theory of Chomsky move into the centre of molecular-genetic research [16]. This is associated with the hope that the methods and formalisms developed by Chomsky will also prove suited to the task of elucidating the structures of the molecular language of genetics. How far this hope will carry, we must wait to see. However, in the most general sense it may be expected that the application of linguistic methods to biology will open up completely new theoretical perspectives.

Like the philosophy of language, biology uses a special concept of semantics. This stands here, as we have already stated, for the plan-like and purpose-directed self-construction of living matter which finally arises in its extreme functional order. Thus, biological semantics could just as well be termed "functional semantics". However, functions can as little be inferred from their underlying structures as the semantics of a linguistic expression can be determined from its syntax. Like the semantics of linguistics expressions, the semantics of the genetic information appear to present an irreducible property of living matter. Notwithstanding, any progress that linguistic research achieves concerning the problem of semantics will be an enormous help for a better understanding of the corresponding problem in biology.

This is also demonstrated by the following example. Even though the language of the genes lacks the logical depth of human language, we can draw a surprising parallel between the two concerning the distinction between "sense" and "meaning". Genetic information is passed down from parent to daughter organism primarily as an instruction, a mere "pre-scription", which is stored in the hereditary molecules. This form of existence can be compared to the meaning of a linguistic expression. However, the genetic information does not take effect until it is expressed in the fertilised ovum, and this expression is accompanied by a permanent re-assessment and re-evaluation of the information. In this process, the context-dependence of the genetic information comes into effect—precisely the property that imparts the "sense" to a word. In accordance with Frege, we can say that the terms "instruction" and "information" have the same meaning but different senses.

3   The General Structure of Language

If we wish to know what the language of Nature is, then we must first look for the general structure of language. However, we cannot solve this problem by simply making an abstraction based on human language, as this has unique features that are due to the unique position of humans as thinking beings. In human language we can pronounce judgements; formulate truths; and express opinions, convictions, wishes and so forth. Moreover, human language has various manifestations at the level of interpersonal communication: there are picture languages, gesture languages, spoken languages and written languages. Therefore, to uncover the general structure of language, we must develop our concept of language from the bottom up. Setting out from the most general considerations possible, we must endeavour to develop an abstract concept of language that reflects human language and at the same time is free from its complex and specific features.

This constraint immediately highlights an almost trivial aspect of language, one that is basic for any process of communication: successful communication is only possible when the partners in communication use a common pool of symbols. Further to that, sequences of such must also be structured according to rules and principles which are known to both the sender and the recipient. Such syntactic structures then form the framework for a symbol language shared by sender and recipient.

However, the syntactic structure must fulfil yet another condition: the sequences of symbols from which a language is built up must have an aperiodic structure. This is because only aperiodic sequences are able to encode sufficiently complex information. This can also be shown to be the case for human language.

Let us consider the language of symbols in more detail. Its character can be illustrated by reference to the communicative processes that take place in living matter. Here, the language is initially instructive in Nature, that is—setting out from genetic information—it sets up the posts that the innumerable elementary processes have to follow. However, the organism is anything but a piece of mechanical clockwork that follows a rigid set of symbols. It is better viewed as an extremely complex system of internal feedback loops, one that has continually to orientate and adjust its internal processes in order to preserve itself and its capacity for reproduction. To do this, it must continually take up new information from its environment (both inside and outside the organism) and evaluate and process this information.

The interactive procedures of expression and continual re-evaluation of information in living matter go beyond mere instruction, insofar as this level of signal-processing takes place with reference to the entire context of the signals. To take account of the full breadth of this aspect of communication, let us look at human language, which is context-dependent to a high degree. The context-dependence is even reflected in the inner structure of a language. And this is because sounds, words, etc. are always present in a complex network of interrelationships. Moreover, this is precisely what makes them elements of a language in the first place.

The "internal" structure of language has been investigated systematically within the framework of structuralism. According to this view, every language is a unique set of interrelated sounds and words and their meanings, and this cannot be reduced to its individual component parts, because each of these parts only acquires meaning within the context of the overall structure. Only through the overall structure can the elements of the language coalesce into a linguistic system; in this system, the elements are demarcated from one another and there is a set of rules that assign a linguistic value to each element.

In the light of such insights into the structure of human language one could toy with the idea of turning these insights on their head, regarding every ordered network of interrelationships as a linguistic structure. Such a radical reversal of the structural view of language is by no means new; it has long stood at the centre of a powerful current of philosophical thought, which is named "structuralism".

Its programme has been expressed by Deleuze in the following way: "In reality there are no structures outside that which is language—although it may be an esoteric or even a non-verbal language. The subconscious only possesses structure insofar as the sub/unconscious speaks and is language. Objects only possess structure insofar as they are considered to speak in a language that is the language of symptoms. Objects themselves only have structure insofar as they hold silent discourse, which is the language of symbols" [5].

Opinions may differ concerning the depth and breadth of the relevance of structuralism for our understanding of the world. Such a verdict will depend not least upon whether one regards structuralism as merely a methodological, if relatively exact, tool for the humanities and social sciences, or whether one accepts it as the ultimate authority for our understanding of the world. As long as structuralism recognises the limits that are placed upon any exact science, it will not claim to embrace the total wealth of reality. Rather, it will follow its own intention of investigating the structures of reality as such, i.e., independently of the forms in which these structures are manifested in reality.

4 The "Language" of Structures

The use of terms such as "information" and "language" in biology is often criticised as a naturalistic fallacy. Both terms, it is claimed, fail to characterise natural objects, being illegitimate projections from the non-scientific area of interpersonal communication and understanding into the area of natural sciences. However, this charge is not even justified in a scientific-historical sense. This is because the concept of information in biology is rooted not in the idea of communication but in that of material instruction. A clear example of this is provided by Schrödinger's elaborations on the structure of chromosomes in the 1940s [15].

The later information theory of Shannon likewise fails to provide any direct link to the forms of interpersonal communication. It is true that Shannon initially called his information-theoretical approach "communication theory"; however, he was interested only in the machines, the technology of message-transfer and the channels along which this proceeds (see above).

Consequently, the concept of information as used in the natural sciences and in engineering is not a "natural" concept, but rather a "structural" one. Let us look at this in more depth by considering the entropy of the source of a message (4). For this purpose, we must first lay open the structure of the idea of entropy. This leads us, as a first step, to the need to appreciate the distinction between two levels of description of a system: its microstate(s) and its macrostate.

In information theory, the macrostate describes the higher-order properties of a sequence of symbols, such as its length n or the size λ of the set of symbols it uses. In contrast, the microstate is one of the λ^n possible ways of ordering the symbols. According to Shannon the quantity of information of such a sequence is given by the number of binary decisions that are needed to select a particular microstate out of the set of all possible microstates. The Shannon information thus represents the "potential" information contained in a macrostate, that is, everything that could possibly be known as a result of having complete knowledge of the entire microstate. In other words: Shannon's measure of information is determined by the accuracy with which the respective micro- and macrostates are defined.

Let us now consider the entropy function of statistical thermodynamics. Here, the macrostate of a material system is given by a number of state functions such as pressure, volume, temperature and the like, while the microstate is given by the exact specification of the position and momentum co-ordinates of all the particles that make up the system (or, put differently, a complete description of its approximately time-independent quantum state). As a rule, such a system contains an unimaginably large number of possible microstates. On the assumption that all the microstates have the same prior probability of being realised, the system will in most cases be in the macrostate that has greatest statistical weight, i.e., the greatest number of microstates. Such a distribution is only very rarely followed by another with a lower statistical weight if, as is usual with atomic or molecular systems, the particle numbers involved are large.

Boltzmann suggested that physical entropy should be understood as a strictly increasing function of the statistical weight of a distribution and the entropy of a macrostate should be set to be proportional to the number of its microstates. The mathematical formulation of this idea leads to the equation

$$S = -k \sum_i p_i \ln p_i. \qquad (11)$$

Here p_i is the probability of realisation of the ith microstate and k is a proportionality factor known as Boltzmann's constant. For thermally isolated systems, all microstates possess the same prior probability $p_i = 1/p$, so that the above equation adopts the form

$$S = k \ln W, \tag{12}$$

where W is the number of possible microstates that make up a macrostate. This number is also called the "thermodynamic probability"; it should however not be confused with the mathematical probability, which always lies in the range $0 \le p \le 1$.

Thus, apart from its sign, the Shannon entropy has the same mathematical structure as the Boltzmann entropy as given by (11). Strictly speaking, informational entropy and thermodynamic entropy are two structurally equivalent partition functions, of which one refers to the distribution of messages and the other the distribution of energy among the various quantum states.

In consequence of the structural aspects described above, it is therefore justifiable to equate information with negentropy. Boltzmann had already recognised this when he laid the statistical foundations of entropy, pointing out that the entropy function is at the same time related to the information that we have about the system [2]. Indeed, for the case of an ideal gas this is immediately obvious: the temperature of such a gas is directly proportional to the mean kinetic energy of the gas molecules and represents an *intensive* quantity—that is, it depends not upon the number of particles for which the mean value is calculated. It necessarily follows from this that there must be another quantity, complementary to temperature, that is proportional to the size or extent of the system. Were this not the case, then our knowledge of the total thermal energy of the system would be incomplete. This *extensive* quantity is the entropy.

The entropy function thus is a measure of the loss of information that arises in statistical physics when one dispenses with exact knowledge of the microstate. In this way, Boltzmann brought into focus—for the first time—the fundamental dependence of scientific statements upon the means and abstractions used to describe them.

The view of information and negentropy as being equivalent furnishes the concept of information not with a naturalistic, but rather with a structural interpretation. This leads to the collapse of the frequently expressed criticism that science wrongly regards information as a natural entity. It now becomes clear that this criticism rests upon a fundamental misunderstanding. In fact, the natural sciences rarely concern themselves directly with natural entities. Of much more importance for a causal understanding of Nature are the relationships between such entities. This becomes especially clear with the example of physical entropy. Although it is true that entropy is an object of scientific discourse that can be described by a physical quantity, it is not a natural entity in the same sense that trees, stones or other material objects are. Rather, the entropy of a system is a statistical partition function, the objective nature of which lies exclusively in the objective nature of the distribution of energy that it expresses.

Information is precisely *not* a "natural" entity in the narrow sense, but rather a "structural" entity that is structurally equivalent to the physical entropy. Consequently, information theory should be assigned neither to the natural sciences nor to the humanities. It belongs to the rapidly increasing branch of the so-called structural sciences. Their programme is to describe the comprehensive, abstract structures of reality, independently of where we encounter them and of whether they characterise living or non-living systems, natural or artificial ones. The structural sciences form a self-contained scientific view of reality, seeking the elements that combine the traditional natural sciences and the humanities.

In general, one obtains the laws of structural sciences by removing all the empirical constants from empirical laws and replacing them with logical constants. The structural laws then have the same syntactic structure as the empirical laws from which they were derived. In science, there are numerous examples of such structural equivalences. A very simple example from physics is the parity of structures of the law of gravity, which describes the attraction between two masses, and Coulomb's law, which describes the force between two charges. One may also mention the Fourier equation, which is applied in electrodynamics, hydrodynamics and thermodynamics. Another well-known example is that of the Lotka–Volterra equations, which are used for the description of dynamic systems not only in biology, but just as much in chemistry and economics.

Above all, the emergence of the structural sciences was due to the investigation of complex systems. As there are no additional natural laws that apply only to complex systems, their underlying laws must be sufficiently abstract to do justice to the vast diversity of complex phenomena, each with its individual characteristics. This explains the spectacular growth of the structural sciences in our day. Alongside information theory, the middle of the last century saw the establishment of cybernetics, game theory, systems theory and semiotics. The classical structural sciences have been enriched in recent decades by disciplines as important as complexity theory, network theory, synergetics and the theory of self-organisation. At the same time, the structural sciences form a mould for possible sciences, comparable with mathematics, which one may regard as a prototype of a structural science. Unlike mathematics, however, the structural sciences develop only through interaction with the experienceable structures of reality. And, most importantly: the structural sciences also contain the key that opens the door to the semantic dimension of information.

5 A Quantitative Approach to the Semantics of Information

Meaningful information in an absolute sense does not exist. Information acquires its meaning only in reference to a recipient. Thus, in order to specify the semantics of information one has to take into account the particular state of the recipient at the moment he receives and evaluates the information. In human communication the recipient's state is insofar a unique one as it is determined by his prior knowledge, prejudices, desires, expectations and so forth. But can those individual and historically determined circumstances ever become the subject of an exact science based upon generalised principles, rules and laws?

Against this background, one must give credit to the efforts of the cultural philosopher Cassirer to bridge over the apparent dichotomy between the particular and the general [3]. The particular, as Cassirer argues, does not become such by being a thing apart from general principles, but rather by entering into these in a relationship of ever-increasing multiplicity. The individual and particular—as we might also say—"crystallises" out of the network of its dependences upon general principles. This is a highly interesting figure of thought, and it points to a way in which one may comprehend the individual and particularly by reference to the general, even though this can never be performed exhaustively.

On the basis of this idea, we will now develop a novel approach to information that finally leads to a quantification of the value and thus the semantics of information. To do this, we must first see what the general rules are that constitute the semantics of information in the sense explained above. According to prevailing opinion, at least four general aspects are decisive: the novelty value, the pragmatic relevance, the selectivity and the complexity of information.

Already in the 1950s, Bar-Hillel and Carnap undertook the attempt to quantify the meaning of information on the basis of philosophical and linguistic considerations [1]. They allowed themselves to be guided by Shannon's idea, according to which the essence of information consists in removing uncertainty. Setting out from this idea, they transferred Shannon's information metric to linguistic statements and assessed the information content of such statements on the basis of how strongly the set of possible expected statements could be restricted by the arrival of a particular statement. However, this concept could only be realised within an artificial language, so that the range of its applicability necessarily remained extremely narrow.

In comparison with this, the idea of measuring semantic information by its pragmatic relevance has been much more fruitful. The pragmatic approach to the semantics of information has been favoured by many authors (see for example [13]). This approach takes account of the fact that meaningful information can be characterised by the reaction which it initiates on the part of its recipient. Such reactions can in turn express themselves in actions taken and thus provide a measurable indicator of the content of the information in question. The concept of the "pragmatic relevance" of a piece of information is by no means restricted to consciously acting recipients. Under certain circumstances the pragmatic relevance of a piece of information is already seen in the functional mosaic of the receiving system, as for example is the case in the living cell when genetic information is retrieved and processed.

This functional aspect of information has been investigated in detail by Eigen in his theory of the self-organisation and evolution of matter [6]. The theory shows that at the level of genetic information the evaluation of the functional content of information takes place through natural selection. Moreover, in this case the semantics of genetic information can even be objectified and quantified by its "selection value". In general, the basic assertion holds: evaluated information is selective, whereby selectivity reveals itself as being a fundamental aspect of semantics.

Another theoretical approach which has been developed by the author is aimed at the prerequisites that have to be fulfilled by syntax if it is to be a carrier of semantic information [11]. One such prerequisite is the vast aperiodicity of the arrangement of the symbols, as only aperiodic sequences possess adequate scope for the encoding of meaningful information. From this point of view, the aperiodicity of the syntax can be considered as a realistic measure for the complexity of the semantics that can be built up on it.

In face of the different approaches to the semantic dimension of information one might obtain the impression that there is a basic contradiction between them. However, this is not the case. On the contrary: the breadth of this variation is a necessary consequence of the idea described above, according to which the semantics of information arise through the multiplicity of its general aspects.

Let us demonstrate this by referring to an example from human language. Consider the statement: "It will rain in Berlin tomorrow". Anyone receiving this message can evaluate it according to various criteria. It can, for example, be evaluated for its novelty. But if a recipient of the message has already heard the weather forecast, then its value for him will at best be that of confirming information already in his possession, and the novelty value will be low.

However, one may also ask what the pragmatic relevance of the message is. This is equally large for all its recipients who are likely to be exposed to the weather in Berlin of the following day. Conversely, it will have no (or only marginal) relevance for those recipients who in any case are planning to spend that day outside Berlin.

Comparable considerations apply to the selectivity of a message. Within the frame of all conceivable weather forecasts, our example is highly selective and, thus, considerably more meaningful than a message stating that tomorrow it will rain somewhere in the world. But even that information is dependent upon its recipient. For a recipient outside Berlin, both messages are presumably equally uninteresting. On the other hand, the selective value of the message will increase with the recipient's closeness to Berlin on the day in question.

Last but not least, the complexity of a piece of information clearly also plays a part in its evaluation. If one reduces the complexity of the message "It will rain in Berlin tomorrow" to the sentence "It will rain in Berlin", then the information content for most recipients will fall because of the message's poor selectivity in respect of the time point referred to.

This example demonstrates that general principles alone do not constitute a sufficient condition for determining the value and thus the semantics of a given item of information. It is rather the specific weighting of these principles by the recipient that gives the information its individual and particular meaning.

In general, the essential characteristics of semantics, such as the novelty value, pragmatic relevance, selectivity, complexity, confirmation and so forth, can be conceived of as elements W_k of a semantic code C_{sem}:

$$C_{sem} = \{W_k\} \qquad (k = 1, 2, \ldots, n). \tag{13}$$

However, in contradistinction to the usual understanding of a code, the semantic code does not possess any rules for assignment or translation. Rather, the elements of the semantic code determine the value scale that a recipient applies to a piece of information received. Strictly speaking, the semantic code is a superposition principle that, by superimposition and specific weighting of its elements, restricts the value that the information has for the recipient and in this way becomes a measure for the meaning of the information.

If the elements W_k have the weights p_k (with $\sum_k p_k = 1$) for the evaluation of a piece of information I_j by the recipient, then an adequate measure for the information value $W(I_j)$ would be a linear combination of the weighted elements W_k :

$$W(I_j) = \sum_k p_{jk} W_k \quad \text{with} \quad \sum_k p_{jk} = 1 \qquad (j = 1, \dots, m). \qquad (14)$$

This measure, in turn, has the same mathematical structure as the entropy of a message source (4). However, in place of the weighted messages, relation (14) contains the weighted values W_k of a chosen message I_j. At the same time, the number k is a measure of the fine structure of the evaluation scale: the greater k is, the sharper, i.e., the more differentiated is the evaluation by the recipient. In the limiting case, where the only value a recipient attaches to a message is its novelty, (14) reduces to the information measure (2) of classical information theory.

The information value $W(I_j)$ is a relative and subjective measure insofar as it depends upon the evaluation criteria of the recipient. However, for all recipients who use the same elements of the semantic code, and who for a given message I_j assign the same weights to these elements, $W(I_j)$ is an objective quantity.

6   The Complexity of Semantic Information

The elements of the semantic code can be—at least in principle—quantified. For novelty value, Shannon's definition of information content is a good example. For complexity, the measure offered by algorithmic information theory seems appropriate. Let us look more closely at the latter, because it allows far-reaching conclusions to be drawn concerning the context-dependence of information. To investigate this, we take up the central thesis of philosophical hermeneutics, i.e., that one can only understand something if one has already understood something else (for this issue see [12]). This thesis we now put onto a quantitative basis. Expressed in the language of information theory, the question would be: how much information is needed in order to understand another piece of (meaningful) information?

An exact answer to this question seems at first glance impossible, as it still contains the problematical concept of "understanding". However, surprisingly, an answer is nonetheless possible, even though one must restrict the consideration to the minimum condition for any understanding. This minimum condition is the obvious one that information which is to be understood must first be registered by the receiver. On the other hand, there are good reasons to assume that a symbol sequence carrying information must have an aperiodic structure (see Sect. 5). For such sequences there is however no algorithm that would make it possible to deduce the rest of the sequence if only a part of it were given. Therefore, the recipient must be in possession of the entire information in question, i.e., of the entire sequence of symbols, before the actual process of understanding can commence. Thus, even the act of registration demands a quantity of information that has at least the same degree of complexity as the sequence of symbols that is to be understood.

Let us take an example. Consider again the information "It will rain in Berlin tomorrow". We are not concerned with its truth content, but only with the complexity of the sequence of letters, which we can at the same time regard as a measure of the complexity of its semantics. In the above sentence, the complexity of the sequence of letters is at a maximum, since there is no algorithm that would be shorter than this sequence and with which at the same time the sequence could be extended or augmented. The structure, i.e., the syntax of this sentence is aperiodic and, in that sense, random. If, in contrast, the sequence were periodic, or largely periodic, then its inherent regularity, or law-like structure, would allow it to be compressed— or, if a part of it were already known, would allow the other part to be generated. This means that, from a syntactic point of view, meaningful letter sequences are always random sequences. However, this statement should not be inverted! Not every random arrangement of letters represents a meaningful sequence.

These conclusions are fundamental. They remain unaffected by the criticism that every language possesses syntactic rules according to which the words of the language are allowed to be assembled into correctly formed sentences. However, such rules only restrict the set of random sequences that can carry meaning at all. They do not allow any converse inference to be made about the meaning content itself. This is because the meaning of a sentence invariably depends upon the unique order of its letters.

One can express this in another way: meaningful information cannot be compressed without loss of some of its meaning. Of course, the content of a piece of information may sometimes be reduced to its bare essentials, as done in telegram style or in boulevard newspapers, but some information is always lost in this process.

Nevertheless, the fact that there is no compact algorithm for the letter sequence discussed above cannot be proved in a strict manner. This is because there might be some simple rule, hidden in the letter sequence, which we have not noticed until now. However, this hypothesis is arbitrarily improbable, as almost all binary sequences are aperiodic, i.e., random.

Even if the randomness of a given individual sequence cannot be proven, one can at least determine the proportion of sequences with—let us say—a complexity of $K = n - 10$ among all combinatorially possible binary sequences of length n. This is done by simply counting the sequences that can generate a sequence of complexity $K = n - 10$.

There are 2^1 sequences of complexity $K = 1$ with this property, 2^2 sequences of complexity $K = 2, \ldots\ldots$ and 2^{n-11} sequences of complexity $K = n - 11$.

The number of all algorithms of complexity $K < n - 10$ thus adds up to

$$\sum_{i=1}^{n-11} 2^i = 2^{n-10} - 2. \tag{15}$$

As no algorithm with $K < n-10$ can generate more than one binary sequence, there are fewer than 2^{n-10} *ordered* binary sequences. These make up one 2^{10}th of all n-membered binary sequences. This means that among all binary sequences of length n only about every thousandth sequence is non-random and possesses a complexity $K < n - 10$.

To summarise the result: in order to understand a piece of information, one invariably needs background information that has at least the same degree of complexity as the information that is to be understood. This is the sought-for answer to the question of how much information is needed to understand some other piece of information. This finding gives the phenomenon of the context-dependence of information and language a highly precise form.

7 Concluding Remark

According to the model of the semantic code developed in this paper, the specific content of meaningful information is constituted by the superposition and the weighting of the general aspects of that information. It is quite conceivable that meaningful information may be assembled according to the same pattern in the brain. Thus, theoretical and experimental research suggests that brain cells, stimulated by sensory impressions, join up into a synchronous oscillating ensemble, whereby the individual information segments, scattered over different parts of the brain, are united into a self-consistent cognitive structure. In the light of the above considerations the synchronisation of oscillatory activity appears as the neuronal expression of a principle that we have identified as the generating principle of semantics. It may be that this principle will prove to be a useful model for the so-called neural code that neurobiology is seeking and which is assumed to be the guiding principle for the constitution of meaningful information in the brain.

References

1. Bar-Hillel, Y., Carnap, R.: Semantic information. Br. J. Philos. Sci. **4**: 147–157 (1953)
2. Boltzmann, L.: Vorlesungen ber Gastheorie, 2 Bd. J. A. Barth, Leipzig (1896/1898)
3. Cassirer, E.: Substanzbegriff und Funktionsbegriff. Wissenschafliche Buchgesellschaft, Darmstadt (1980)
4. Chaitin, G.J.: On the length of programs for computing finite binary sequences. J. Assoc. Comput. Mach. **13**, 547 (1966)

5. Deleuze, G.: Woran erkennt man den Strukturalismus?, p. 8. Merve, Berlin (author's translation) (1992)
6. Eigen, M.: Selforganisation of matter and the evolution of biological macromolecules.
   Naturwissenschaften **58**, 465–523 (1971)
7. Hartley, R.V.L.: Transmission of information. Bell Syst. Tech. J. **7**, 535–563 (1928)
8. Kolmogorov, A.N.: Three approaches to the qunatitative definition of information. Problemy
   Peredachi Informatsii **1**, 3–11 (1965)
9. Küppers, B.-O.: Information and the Origin of Life. MIT, New York (1990)
10. Küppers, B.-O.: The context-dependence of biological information. In: Kornwachs, K.,
    Jacoby, K. (eds.) Information. New Questions to a Multidisciplinary Concept, pp. 135–145.
    Akademie Verlag, Berlin (1995)
11. Küppers, B.-O.: Der semantische Aspekt von Information und seine evolutionsbiologische
    Bedeutung. Nova Acta Leopoldina **294**, 195–219 (1996)
12. Küppers, B.-O.: Information and communication in living matter. In: Davies, P.,
    Gregersen, N.H. (eds.) Information and the Nature of Reality, pp. 170–184. Cambridge
    University Press, Cambridge (2010)
13. McKay, D.: Information, Mechanism and Meaning. MIT, New York (1969)
14. Rényi, A.: Probability Theory. Elsevier, Amsterdam (1970)
15. Schrödinger, E.: What is Life?. Cambridge University Press, Cambridge (1944)
16. Searls, D.B.: The language of genes. Nature **420**, 211–217 (2002)
17. Shannon, C.E.: A mathematical theory of communication. Bell Syst. Tech. J. **27**, 379–423,
    623–656 (1948)
18. Solomonoff, R.J.: A formal theory of inductive inference. Inform. Contrl. **7**, 1–22, 224–254
    (1964)
19. Weaver, W., Shannon, C.E.: The Mathematical Theory of Communication, p. 8. University of
    Illinois Press, London (1963)
20. von Weizsäcker, C.F.: The Unity of Nature. Farrar Straus Giroux, New York (1981)

Talking About Structures

Rüdiger Inhetveen and Bernhard Schiemann

> *There has been a great deal of speculation in traditional philosophy which might have been avoided if the importance of structure, and the difficulty of getting behind it, had been realised.*
>
> Bertrand Russell

Abstract In the first part, we try to give a sufficiently precise definition of the concept of "structure" as a term that denotes abstract entities. Some examples are included. The second part is devoted to a short analysis of how the concept is used in different scientific areas, reaching from mathematics and physics to the humanities and computer science.

1 Introduction

During the second half of the past century a series of books were published containing the term "structure" in their title. Let us only mention the following:

- *Structure and Appearance* (1961) by Nelson Goodman
- *Strukturwandel der Öffentlichkeit* (1962) by Jürgen Habermas
- *The Structure of Scientific Revolutions* (1962) by Thomas Kuhn
- *Erfahrung und Struktur* (1968) by Friedrich Kambartel
- *La struttura assente* (1968) by Umberto Eco
- *The Logical Structure of Mathematical Physics* (1971) by Joseph D. Sneed

R. Inhetveen (✉) · B. Schiemann
Former Chair for Artificial Intelligence and Center for Ethics and Scientific Communication, Friedrich-Alexander-Universität Erlangen-Nürnberg Erlangen, Germany
e-mail: inhetveen@informatik.uni-erlangen.de; bernhard.schiemann@web.de

B.-O. Küppers et al. (eds.), *Evolution of Semantic Systems*,
DOI 10.1007/978-3-642-34997-3_5, © Springer-Verlag Berlin Heidelberg 2013

- *Structure and Evolution of the Universe* (1975) by Ya. B. Zel'dovich and I. D. Novikov
- *Structure and Interpretation of Computer Programs* (1985) by Harold Abelson, Gerald Sussman, and Julie Sussman.

This list is far from being complete, but it clearly indicates that it had become modern to talk of structures within very different fields of scientific research reaching from epistomology to cosmogology, from social sciences to semiotics, from theoretical physics to computer science. In what follows we shall look first on the concept of structure, second on some fields of science in which it is used and finally we shall try to come to some methodological conclusions.

2   The Concept of Structure

Let us begin with a look at the word "structure" and its use in science and philosophy. In the English language the word "structure" was first[1] used in the beginning of the seventeenth century in connection with buildings. It gradually became used metaphorically and reached science at the end of the nineteenth century. In 1919 the mathematician and philosopher Betrand Russell (cf. [14], p. 60f.) was the first to give an explicit definition of this notion, and it was based on *abstraction*. In 1928 this definition was also given by Rudolf Carnap in his book *Der logische Aufbau der Welt* (which was translated into English under the title *The logical structure of the world*) [5]. Let us first recall Russell's definition. In his *Introduction to Mathematical Philosophy* he states [14, p. 61] he calls "structure"

> a word which, important as it is, is never (so far as we know) defined in precise terms by those who use it.

and on page 53/54 he writes:

> We may define two relations P and Q as "similar," or as having "likeness," when there is a one-one relation S whose domain is the field of P and whose converse domain is the field of Q, and which is such that, if one term has the relation P to another, the correlate of the one has the relation Q to the correlate of the other, and vice versa. A figure will make this clearer. Let x and y be two terms having the relation P. Then there are to be two terms z, w, such that x has the relation S to z, y has the relation S to w, and z has the relation Q to w.

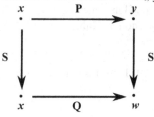

> If this happens with every pair of terms such as x and y, and if the converse happens with every pair of terms such as z and w, it is clear that for every instance in which the relation P holds there is a corresponding instance in which the relation Q holds, and *vice versa*; and this is what we desire to secure by our definition.

---

[1] According to *An etymological dictionary of the English language* by Walter W. Skeat, Oxford [4]1910 [16].

The diagram shown in this definition has a property nowadays called *commutative* and shortly written as

$$Q \circ S = S \circ P.$$

The modern reader of Russell's text will recognize in it a definition of *structure* by abstraction: the first step being a definition of structural equality or—in Russell's terms: "likeness" or " similarity"—and the second step being a restriction of the language used when talking on structures to terms that are *invariant* with respect to the equivalence relation in question. In this way structures are abstract objects represented by "concrete" relations; or—as mathematicians will say—structures are equivalence classes of relations with respect to the equivalence relation called "likeness".

All that means that in order to define a structure we need an "instance" first, that is: a set (Russell used the term "field"), one or more relations defined on it, and a kind of list naming all pairs (or triples) of elements having the relation in question. Thus the problem of representing structures arises.

In simple cases it is indeed a finite list that solves our problem. And in the most simple case we have just one object (to have an example: a person **P**) and one relation connecting this object with itself (in our example: **P**'s knowing (k) himself. Thus our "list" consists of a single line telling us that $k(\mathbf{P},\mathbf{P})$. This can be visualized as

P $\circlearrowright$ k

Forgetting now the example and talking abstract we get something we may call *the singleton structure*, represented by the following graph:

Note the emphasis on the definite article: abstract objects are *individuals*. The next case is mentioned here only to have a name for it in our later considerations: we have two objects (example: Socrates, **S**, and Plato, **P**), with the relation "being teacher of", visualized as

$$\mathbf{S} \xrightarrow{\ t\ } \mathbf{P}$$

Let us call the *abstract structure*, represented by

in what follows *the distinction structure*,[2] because it occurs whenever a distinction is made between different things, as for example [7, p. 29], where Greimas gives his "second definition of structure"[3] saying

---

[2]With a symmetric relation "is different from", so the arrowhead is suppressed.

[3]"[...] the first definition, one generally employed, of the concept of structure: the presence of two terms and the relationship between them." [7, p. 19].

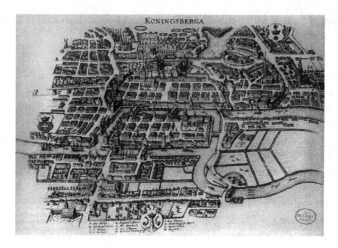

Fig. 1 The bridges of Königsberg

structure is the mode of existence of signification, characterized by the presence of the articulated relationship between two semes;

or in the famous "semiotic square" (see e.g. [19, p. 72]):

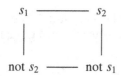

with "opposite terms" s_1 and s_2 as, e.g., "black" and "white".

To give at least one less simplistic example let's consider what might be called the *Euler-structure*: The name is given in honor of Leonhard Euler, who demonstrated the unsolvability of the famous problem of the bridges in Königsberg (cf. Fig. 1).

Looking at this view of Königsberg[4] we see seven bridges connecting four parts of the city. The problem was to make a walk crossing each bridge exactly once. Instead of wasting time by walking Euler used the view to construct a "concrete" structure by making the four parts of the city the "objects" and the seven bridges the relations ("connected by a bridge") between them. Instead of a list he used a graph that looked like the one here depicted (cf. Fig. 2).

The problem of the bridges would only have a solution if each node in this structure would have an even number of lines leaving it; but none of them has. So the problem has no solution. The example shows that structures are not "found" in "nature" but have to be adequately "invented" and "projected" into a real situation.

---

[4]Picture taken from the "Königsberg"-article in `wikipedia.de`.

Fig. 2 The Euler-graph

$$\ldots \bullet \overset{n}{\to} \overset{n+1}{\bullet} \ldots$$

Fig. 3 An order structure

In computer science we are accustomed to the lists mentioned above, as an example from the LISP programming language will illustrate:

```
(defstruct <structure-type>
    (<field1> <default-value1>)
    (<field2> <default-value2>)
    (<field3> <default-value3>)
    ...)
```

Structure-types in LISP

At first sight there are no relations visible: they are hidden in the innocent term "structure-type", as a "real" example will show:

```
(defstruct customer
    (name Miller)
    (first-name John)
    (assets 1,000,000-$)
        ...)
```

Structures in LISP

Being a customer of course means being a customer of a business company, so it *is* a relation. The example will also stand for the use of lists in database systems— and the mere name of "entity-relationship-model" tells us that they are concerned with structures all the time.

In more complex situations, especially when the "field" of elements considered is non-finite, there is no chance to give a list of all pairs of elements connected by the relation(s) defined on it. Sometimes the case remains simple because it is possible to define a *partial structure* which is repeated indefinitely often in the structure at hand. This situation we meet in the ordered set of natural numbers with the partial structure leading from n to its successor $n + 1$ (Fig. 3).

Fig. 4 The structure of table
salt

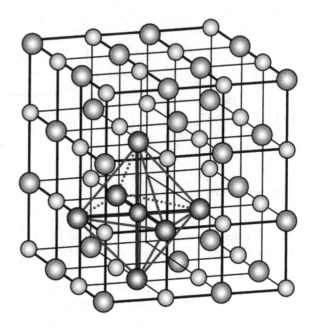

Fig. 5 A partial structure

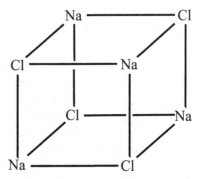

As a little more complex example we may take ordinary table salt (cf. Fig. 4)[5]
where a cube consisting of chlorin and sodium ions is representing this partial
structure (cf. Fig. 5).

Let us emphasize again that table salt does *not* look like shown in these pictures;
instead *we* are making a description using spatial images of what in fact is invisible:
the ions as well as the chemical forces connecting them.

In even more complex cases where this kind of visualization fails, mathematics
has given us the means to master this situation too: we can (sometimes) describe
such structures *axiomatically*. The best known example here is of course euclidean
geometry, where David Hilbert in his *Grundlagen der Geometrie* (1899) [9]

---

[5]Picture taken from: wikipedia.org/wiki/Bild:NaCl-Ionengitter.png

delivered a complete description of the structure behind the theory. But this was not clearly expressed in the text—mainly because Hilbert used the pictorial terms of Euclid's geometry instead of "dummy" words for the objects and relations constituting the structure at hand. Only in 1942 Heinrich Scholz delivered a "tranlation" of Hilbert's text into a purely structural one, transforming for example Hilbert's axiom

> I.1 *For two points A, B there always exists a straight line a which contains each of the two points A, B.*

into the phrase (see [15, p. 283]; my translation, R.I)

> *For two A-things P_1, P_2 there exists always a B-thing G such that P_1 and P_2 are in R_1-relation to G.*

3   The Use of Structures

Mathematics. With the last example we have entered the realm of mathematics which nowadays is "the" science of structures. That will say mathematics *deals* with structures. So there must of course be a clear definition of the concept—and there is. In full generality it can be found in the monumental work of the Bourbaki-group, to be precise: in chapter four of the volume devoted to set theory. In the booklet of results a less rigorous definition is given, with a note reading (see [4, p. 52, note (∗)]; my translation, R.I.):

> The reader will note that the indications given in this paragraph remain rather vague; they are here only for heuristic reasons and it seems nearly impossible to give the general and precise definitions concerning structures outside the frame of formal mathematics (see *Set Theory*, chap. VI).[6]

That is why we shall not repeat this definition here. In fact it differs from Russel's definition not in essence but in generality. Instead of going into technical details we therefore shall give a short sketch of what is done with structures in mathematics, once they are defined.

First, three *basic sorts* of structures are distinguished: order structures (as, e.g., the natural numbers with their "natural" order), algebraic structures (as, e.g., the ring of the integers), and topological structures (as, e.g., the common euclidean space with its metric). Second, operations are defined to gain new structures from these by *canonical* methods: substructures, quotient structures, product structures, and some more. These methods are *common* to all types of structures. Third—and most important—*multiple structures* are defined, that is: sets carrying two or

---

[6]Original text: Le lecteur notera que les indications données dans ce paragraphe restent assez vague; elles ne sont là qu'à titre heuristique et il ne semble guère possible d'énoncer des définitions générales et précices concernant les structures, en dehors du cadre de la Mathématique formelle (voir *Ens.*, chap. IV).

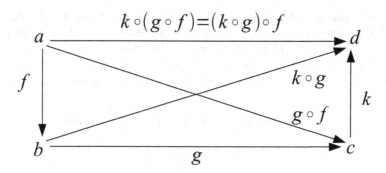

$$k \circ (g \circ f) = (k \circ g) \circ f$$

Fig. 6 Structural compatibility

> **Wir machen uns innere Scheinbilder oder Symbole der äußeren Gegenstände, und zwar machen wir sie von solcher Art, daß die denknotwendigen Folgen der Bilder stets wieder die Bilder seien von den naturnotwendigen Folgen der abgebildeten Gegenstände.**

Fig. 7 Hertz "Die Prinzipien der Mechanik in neuem Zusammenhange dargestellt"

more different types of structures: the most prominent example being the reals as an ordered field with a "natural" metric.

A further central idea using multiple structures consists in *compatibility* of the two or more structures at hand. As an example of what is meant the *associative law* may serve, saying that the upper right and the upper left triangle in the following diagram are commuting (Fig. 6):

The main advantage of studying structures in place of examples is—of course—the generality of results. In view of the abundance of groups it is indeed a deep theorem (first proved by Cayley) that every group is isomorphic to a group of permutations. It illustrates at the same time what are the two main tasks in mathematics as a science of structures: to give really structural theorems and to give manageable *standard representations* of the structures studied.

Historians of mathematics, as, e.g., Dieudonné (see [6, p. 59]), trace this kind of thinking in structures back to the famous *Erlanger Programm* by Felix Klein [11], where a geometry is characterized by a group and the objects studied are the *invariants* under the operations of this group. Here we meet again the idea of invariance mentioned at the beginning that is unavoidable whenever abstraction is being used.

Physics. Let us now look at physics which means *theoretical* physics here. One of the great figures of this discipline, Heinrich Hertz, wrote in his *Die Prinzipien der Mechanik in neuem Zusammenhange dargestellt* of 1894 [8] on the first page of his introduction (Fig. 7):

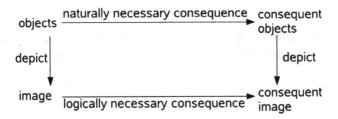

Fig. 8 Equality of structures in physics

> We make us inner images or symbols of the outer objects, and we make them such, that the logically necessary consequences of the images always are the images of the naturally necessary consequences of the depicted objects.

Transforming this thought into a diagram, we get Fig. 8, where the diagram is meant to be commutative, that is: the two relations of "consequence" are "similar" in Russell's terms.

Now it is not difficult to pronounce what Hertz has said into the language of structures, getting the result that the structure of the outer objects (i.e., the real world) has the same structure as the "inner images" we make of them. In short: theory and reality are structurally isomorphic.

So in physics too we find structural thinking towards the end of the nineteenth century. Nowadays physicists are more modest in their assertions on "reality". They talk of *models* (in place of "inner images") which they try to make successful for *descriptions* of the outer world, avoiding ontological commitments. But structural thinking remains in use as Hertz had suggested. The way of doing it consists in an extensive use not just of mathematics but also of structural mathematics.

One result of this way to work is that physics itself has become a *structural science*—though it was not explicitly mentioned in the long list of structural sciences presented by Professor Küppers [12]. To see how this is to be understood let us start from what has been called "structural laws". We do not maintain that the examples given (law of gases respectively of supply and demand; entropy in thermodynamics and in information theory) are not good candidates for structural laws or conceptions. Instead we would like to add two more candidates for structural laws.

The first example remains on the level of abstraction known from Küppers's examples; it's the law of "harmonic oscillation":

$$\ddot{x} + c \cdot x = 0 \tag{1}$$

Taken as a formula, it belongs to mathematics: it's an easy to solve ordinary second degree differential equation. But it becomes a law in physics under *interpretation* (of the domain of the variable, of the dots indicating differentiation with respect to time, and of the constant factor c). And according to this interpretation it governs phenomena of Newtonian mechanics as well as of electric circuits, different regions described by different theories. Exactly this is meant when saying it is a structural law.

Lex. II.

Mutationem motus proportionalem esse vi motrici impressæ, & fieri se-
cundum lineam rectam qua vis illa imprimitur.

Fig. 9 Newton's second "law"

The more interesting example, however, remains entirely within mechanics.
It is the famous "second law" of Newton's *Philosophiae naturalis principia*
mathematica (1678) (cf. Fig. 9, [13][7]), telling us that the change of motion (today
called *acceleration*) is proportional to the moving force.

In modern notation this reads:

$$\mathbf{F} = m \cdot \mathbf{a} \tag{2}$$

Here the "acceleration" $\mathbf{a}$ is a quantity that can be measured directly, and the
same holds for the factor that turns proportionality into equality, that is, for the
"mass" m. But $\mathbf{F}$ is a sophisticated entity: it has been called a "theoretical term",[8]
thus indicating that a direct measurement is impossible: each measuring device must
already make use of a law, e.g., Hooke's law, to work properly. Therefore Newton's
second "law" sometimes was read as a definition of mass.

A structural look on the above formula gives another picture: the letter $\mathbf{F}$ is used
as a symbol *relating* Newton's "law" to at least another law containing the same
symbol for a *special case* of force, as e.g.:

$$\mathbf{F} = -k \cdot \mathbf{x} \quad \text{(HOOKE's law)} \tag{3}$$

$$\mathbf{F} = \gamma \frac{m_1 m_2}{r^2} \frac{r}{|\mathbf{r}|} \quad \text{(NEWTON's law of gravitation)} \tag{4}$$

$$\mathbf{F} = -e_0 \mu \mathbf{E} \times \mathbf{B} \quad \text{(LORENTZ's law for electric charge in magnetic fields)} \tag{5}$$

and many more. Setting one of these terms equal to the $\mathbf{F}$ of Newton's law we get
a differential equation whose solution, with "fitting" initial or boundary conditions,
is called an *explanation* of the corresponding phenomen on (mechanical oscillation,
planetary movement, trajectories of charged particles in magnetic fields). Newtonian
mechanics in this sense supplies us with what is called the *structure* of an
explanation. It is itself called a structural science exactly for this reason.

In the ongoing efforts of establishing so-called "unified theories" for the four
fundamental interactional forces of physics we see another example of structural
thinking. The final result, a GUT (grand unified theory), is not yet reached, but,
interestingly, we "know" some of its formal—and that means: structural—properties

---

[7]The quotation is from the third edition (1871), p. 13.
[8]By Sneed (1971) [17], for example.

(Langrangian formalism, symmetry properties connected to conservation laws according to Emmy Noether's theorem, and perhaps some more). Developments of this kind are the background for people writing books on "the structure of a physical theory".

Humanities. Let us finally have a short glimpse on structural thinking in the humanities. It seemed a breakthrough in the 1960s of the last century when "structuralism" was invented and discovered as a tool for cultural studies (Levi-Strauss), analysis of literature (Umberto Eco), and—of course—semantics (Algirdas-Julien Greimas). If we look for the concept of structure where this title was derived from, we seldom find explanations, and if so, they are not very convincing, as the already quoted example of Greimas (see [7, p. 29]) shows: "[...] the first definition, one generally employed, of the concept of structure: the presence of two terms and the relationship between them." The concept is used implicitly or in a purely colloquial way. In the best cases it indicated that some kind of abstraction was performed. But even then it remained open with respect to which equivalence relation. Of course the authors talked of relations but, alas, relations we meet everywhere we look. So it is not surprising that towards the end of that decade critical positions were published. We shall quote only one here, coming from the French social scientist Raymond Boudon. As early as 1968 he published a small booklet entitled *What is the notion of "structure" good for? Essay on the signification of the notion of structure in the humanities* (see [3]). At the end of a thorough study of "structural" works in his field he concludes on page 213:

> If by "structural method" we understand a set of procedures which allow to get, with respect to any object, a theory on the level of verifiability and as elaborated as possible, which allows to explain the constitutive elements of that object, then we can affirm that such a method does not exist

and several years later, in 1974, his colleague Louis Althusser added [1, p. 57]:

> [the] "flirt" with structuralist terminology has surely exceeded the allowed measure.

4 And Computer Science?

As indicated near the beginning of this talk computer science is using structures in different places and hopefully in a better way than just seen for the humanities. Everyone would agree that a huge amount of, e.g., sales statistics of big companies, need structures to store them in a controlled way. The success of computer science building structures is on the one side related to the terms database (DB), Entity Relationship model (ER), and the Structured Query Language. If one would measure success of this structure technique, just take the amount of money the second biggest enterprize software company Oracle can make with it.

On the other side techniques motivated by the possibilities given by computer reasoned ontologies lead to program structures that are able to model the real-

Fig. 10 Structural equality in MAS

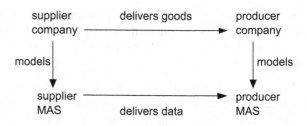

ity directly. In the DFG sponsored program "Intelligente Softwareagenten und betriebswirtschaftliche Anwendungsszenarien" we[9] build a multiagent system that implements the supply chain event management and models the relation between companies in terms of being a supplier or a producer, etc. Please recall Fig. 8 when looking at Fig. 10. To understand the business system details please read [2]. What Fig. 10 wants to illustrate is that for each company a corresponding model in the multiagent system (MAS) is built. The agents in this MAS communicate on facts that are formulated according to the enterprize ontology (see [10, 18]) in DAML+OIL, OWL DL, respectively. This zoom into this special application shows that not only the program or its parts are structured, the interaction/communication is structured as well: the interaction steps in protocols (like FIPA ACL conformant speech act pairs) and the content of the messages structured according to formal ontologies.

References

1. Althusser, L.: Eléments d'autocritique. Hachette, Paris (1974)
2. Bodendorf, F., Zimmermann, R.: Proactive supply-chain event management with agent technology. Int. J. Electron. Commerce **9**(4), 58–89 (2005)
3. Boudon, R.: A qoui sert la notion de "Structure"? Essai sur la signification de la notion de structure dans les sciences humaines. Éditions Gallimard, Paris (1968), (= vol. cxxxvi de la collection "Les Essais")
4. Bourbaki, N.: Théorie des ensembles. Fascicule des resultats, Chapter 4. Hermann, Paris (1964)
5. Carnap, R.: Der logische Aufbau der Welt. Ullstein, Frankfurt/M., Berlin, Wien (1979); Chapter 1, 1928; Chapter 4, 1974, Text according to the unchanged Chapter 4, 1974 edition
6. Dieudonné, J.A.: Geschichte der Mathematik 1700–1900. Ein Abriß. Friedr. Vieweg & Sohn, Braunschweig/Wiesbaden (1985) (Original: Hermann, Paris, 1978, in two volumes entitled Abrégé d'histoire des mathématiques 1700–1900)
7. Greimas, A.-J.: Structural Semantics. An Attempt at a Method. University of Nebraska Press, Lincoln and London (1983), with an introduction by Ronald Schleifer. Original: Sémantique structurale: Recherche de méthode, Paris (Larousse) 1966
8. Hertz, H.: Die Prinzipien der Mechanik in neuem Zusammenhange dargestellt. J.A. Barth, Leipzig (1894)

---

[9]University of Erlangen-Nuremberg, Department of Information Systems II and the Chair for Artificial Intelligence.

9. Hilbert, D.: Grundlagen der Geometrie. Teubner, Leipzig (1899)
10. Huber, A., Görz, G., Zimmermann, R., Käs, S., Butscher, R., Bodendorf, F.: Design and usage of an ontology for supply chain monitoring
11. Klein, F.: Vergleichende Betrachtungen über neuere geometrische Forschungen (1872). Reprint in Mathematische Annalen **43**, 63–100, (1893) at http://www.DigiZeitschriften.de
12. Küppers, B.-O.: Elements of a semantic code. In: Evolution of Semantic Systems, pp. XX–YY. Springer, Berlin (2013)
13. Newton, I.: Philosophiae Naturalis Principia Mathematica. London (1678); vol. 3 (1871)
14. Russell, B.: Introduction to Mathematical Philosophy, 2nd edn. George Allen & Unwin, Ltd., London; The MacMillan Co., New York (1920)
15. Scholz, H.: David Hilbert, der Altmeister der mathematischen Grundlagenforschung. In: Hermes, J.R.H., Kambartel, F. (eds.) Mathesis Universalis, Abhandlungen zur Philosophie als Strenger Wissenschaft, pp. 279–290. Benno Schwabe & Co., Basel/Stuttgart (1961)
16. Skeat, W.W.: An Etymological Dictionnary of the English Language, vol. 4. Clarendon, Oxford (1910)
17. Sneed, J.D.: The Logical Structure of Mathematical Physics. Reidel, Dordrecht (1971)
18. Uschold, M., King, M., Moralee, S., Zorgios, Y.: The enterprise ontology. At http://www.aiai. ed.ac.uk/~entprise/enterprise/ontology.html (1995)
19. Volli, U.: Semiotik. Eine Einführung in ihre Grundbegriffe. A. Franke Verlag, Tübingen und Basel (2002), (Original edition as Manuale di semiotica, Rome 2000)

Toward a Formal Theory of Information Structure

Jerry R. Hobbs and Rutu Mulkar-Mehta

Abstract The aim of this chapter is to present a formal theory of the structure of information that will support a variety of statements about documents in various media, their internal structure, and how they function in the world at large. We begin by sketching an approach to anchoring symbolic systems in human cognition and discuss various levels of intentionality that occur. We then consider compositionality in different symbolic systems and the sometimes complex coreference relations that arise from that. This theory is the basis of a program for translating natural language into logical form, and this is described. Then the theory is applied to the specific case of diagrams as information-bearing objects, and a logical theory of Gantt charts is constructed as an illustration. Finally there is a discussion of issues raised with respect to various modalities and various manifestations of symbolic artifacts.

1 Introduction

Search engines today are very good at finding information in text resources. But very often the best answer to a question is in a diagram, a map, a photograph, or a video. For example, consider the questions

What is the Krebs cycle?
How has the average height of adult American males varied over the last
100 years?
How did the Native Americans get to America?

J.R. Hobbs (✉)
Information Sciences Institute, University of Southern Caifornia, 4676 Admiralty Way, Marina del Rey, CA 90292, USA
e-mail: hobbs@isi.edu

R. Mulkar-Mehta
Precyse Advanced Technologies, 1750 Founders Parkway, Suite 154, Alpharetta, GA 30009
e-mail: me@rutumulkar.com

B.-O. Küppers et al. (eds.), *Evolution of Semantic Systems*,
DOI 10.1007/978-3-642-34997-3_6, © Springer-Verlag Berlin Heidelberg 2013

What does Silvio Berlusconi look like?
What happened on September 11, 2001?
When will the various tasks on this project be completed?

The answer to the first should be a process diagram, the second a graph, the third a map with routes indicated, and the fourth a photograph. For the fifth, a news video clip might be the best answer. The answer to the last might best be presented in a Gantt chart.

Search engines are very much poorer at finding this kind of information, and generally they do so by looking at the associated text. We would like to have this kind of information encoded in a fashion that makes it more retrievable, in part by describing it better, and in part by expressing "coreference" relations among material presented in different media. For example, a diagram may bear a kind of coreference relation with a text segment in the document it is a part of. There may be a coreference relation between an object in a photograph and a noun phrase in the caption. In a news video clip we have analyzed, a woman refers to sending her "four children" to local schools, as she is standing by a wall covered with family photographs; there is also a coreference relation here.

The aim of this paper is to present a formal theory of the structure of information that will support a variety of statements about documents in various media, their internal structure, and how they function in the world at large. We begin by sketching an approach to anchoring symbolic systems in human cognition and discuss various levels of intentionality that occur. We then consider compositionality in different symbolic systems and the sometimes complex coreference relations that arise from that. This theory is the basis of a program for translating natural language into logical form, and this is described. Then the theory is applied to the specific case of diagrams as information-bearing objects, and a logical theory of Gantt charts is constructed as an illustration. Finally there is a discussion of issues raised with respect to various modalities and various manifestations of symbolic artifacts.

Thus, Sects. 2 and 3 on Gricean nonnatural meaning capture how single words or other atomic symbols convey information. Sections 4, 6, and 7 describe how multiple words or other symbols combine to convey more complex information.

Symbol systems evolved. One of the key tasks in explaining the evolution of something is hypothesizing incremental steps in its development where each new advancement confers an advantage. Hobbs [12] describes just such a plausible sequence of steps for Gricean nonnatural meaning and natural language syntax. Here we will sketch this account very briefly.

2 Grounding Symbols in Cognition

In this paper we will assume that we have a coherent notion of causality, as in [11], and change of state, as in [15], as well as a theory of commonsense psychology at least rich enough to account for perception, planning and intentional behavior, and what we here call "cognizing," that is, taking some cognitive stance toward, such as belief, thinking of, wondering about, and so on. We will refer to the contents of thoughts and beliefs as "concepts," a general notion that not only subsumes propositions [5] but also includes nonpropositional concepts like "dog" and "near," images, vague feelings of apprehension, and so on. We will assume the "ontologically promiscuous" notation of Hobbs [7], but for typographical convenience, we will abuse it by using propositions as arguments of other propositions, where a proper treatment would reify the corresponding eventualities and use those as arguments. Some of the inferences below are defeasible, and thus the underlying logic must support a treatment of defeasible inference. There are many frameworks for this, e.g., McCarthy [19] and Hobbs et al. [14]. To minimize notational complexity, defeasibility is not made explicit in the axioms in this paper.

The basic pattern that symbols rest on is the perception of some external stimulus causing an agent to cognize a concept.

$$cause(perceive(a, x), cognize(a, c)) \tag{1}$$

where a is an agent, x is some entity, and c is a concept. x can be any kind of perceptible entity, including physical objects and physical properties, states, events, and processes, and, as we will see later, more abstract entities as well. That is, we can perceive a ball, its roundness, and the event of someone throwing it. Among the states that can be perceived are absences. Seeing that someone's car is not in his garage can cause someone to believe he is not at home. Silence, or absence of speech, can often carry very significant meaning.

The concept c may often be hard to put into words, such as the concepts triggered by music or an aesthetic design.

This pattern covers the case of a cloud reminding someone of a dog, where there is no external causal connection between the stimulus and the concept, and the case of smoke making one think of fire, where there is a causal connection, and the intermediate case of an association that has been established by practice, as in a dinner bell making one think of food.

Some concepts are tied in such a way to the entity perceived that they can be called the "concept of" the entity. We could introduce *conceptOf* as a function mapping from the entity to the concept, but since the predicate *cognize* always takes a concept as its second argument, it is simpler to build the coercion into the predicate *cognize*. Thus, if e is an entity, $cognize(a, e)$ says that agent a cognizes the concept of e. The key relation between entities and their concepts is that perceiving the entity causes the agent to cognize the concept of the entity.

$$cause(perceive(a, e), cognize(a, e)) \tag{2}$$

It is important to note, however, that perception can trigger many concepts and that not everything that is cognized needs to be what is perceived. Perceiving a bell can cause an agent to cognize food (as well as the bell). This makes symbols possible.

Where the concept cognized is propositional, we could talk about its truth in the world. That is, it is not only true that e is cognized, but it also holds in the real world–$holds(e, w)$. However, this will play no role in this paper. The meanings of symbols will be strictly in terms of the concepts they invoke in the recipient.

Communication begins when another agent presents an entity causing the first agent to perceive it.

$$cause(present(b, x, a), perceive(a, x)) \qquad\qquad (3)$$

For an agent b to present something to a is for b to cause it to be within the range of a's perception, and this causes a to perceive it.

The recipient agent a must of course be capable of cognition. A greater range of sending agents b is possible. A car that beeps when you don't fasten your seat belt is an agent b that is presenting a signal x for the driver to cognize. It is also possible for collectives to be the sending agent, as in jointly authored documents such as the Constitution of the United States. The agents may or may not exhibit intentionality. Humans do, as do organizations of humans, whereas simple artifacts merely reflect the intentionality of their designer. Sufficiently complex artifacts may exhibit intentionality.

Causality is defeasibly transitive, so Rules (1) and (3) can be combined into the defeasible causal pattern for appropriate x's and c's:

$$cause(present(b, x, a), cognize(a, c)) \qquad\qquad (4)$$

That is, if b presents x to a, it will cause a to cognize the corresponding concept c. For example, a car beeps and that causes the driver to hear the beep; hearing the beep causes the driver to remember to fasten his seat belt. So the beep reminds the driver to fasten his seat belt.

We will refer to the entity presented (x) as the symbol and to the concept evoked (c) as the meaning of the symbol.

A car that beeps cannot be said to have beliefs. Monkeys that emit alarm cries at the sight of a snake or a leopard may or may not be usefully described as having beliefs about the threat. Higher primates probably do have such beliefs, and humans certainly do.

Belief introduces another level of complexity to meaning. Someone sees a fire and runs to warn his friends. The friends don't see the fire themselves, but they interpret his presentation of the alarm as caused by his belief that there is fire, and this causes them to believe it. Belief acts as a kind of carrier of information across space and time.

A theory of belief is useful in social animals for independent reasons, but once they have such a theory, it can enrich their communication. It allows them to reason in formula (4) about why b presented this information, normally, because b believes c, then to reason about why b came to believe this, and then to assess whether they ought to believe it too. A theory of belief allows agents to interpret the content of utterances as mistakes.

3   Intention and Convention in Communication

Presentation by an agent can involve several levels of intentionality, and the perception can involve several levels of recognition of intentionality. First, the presentation can be entirely unintentional, as, for example, when someone conveys their nervousness by fidgeting or shaking their leg. In an abductive account of intelligent agents, an agent a interprets the environment by telling the most plausible causal story for the observables in it. Here a knows nervousness causes fidgeting and the most plausible causal story is that b's fidgeting is because b is nervous. When b says "ouch" and a infers that b feels pain, the account is exactly the same.

When the presentation is intentional, the presenter's goal is to cause the perceiver to cognize something. The presenter's intention need not be recognized. For example, a professor may keep the door to his office closed to lead students to believe he is not in, without wanting them to recognize his intention to communicate that.

Intention is recognized when it is part of an observer's explanation that an event occurs because the agent of the event had the goal that it occur. Defeasibly, agents do what they want to, when they can.

$$goal(g, b) \wedge executable(g, b) \supset cause(goal(g, b), g) \tag{5}$$

(All axioms are universally quantified on the variables in the antecedents of the highest-level implication). If g is a goal of b's and is executable by b (or achievable by an executable action), then its being a goal will cause it to actually occur. We won't explicate *executable* here, but it means that g is (achievable by) an action of which b is the agent, and all the preconditions for the action are satisfied.

When an observer a uses this causal rule, he is recognizing the intention that lies behind the occurrence of the event.

It is most common in human communication that the intention is recognized. Agent b knows that presenting x causes a to perceive x, which causes a cognize concept c. b has the goal that a cognize c. So that causes b to present x. Agent a comes up with exactly this causal explanation of b's action of presentation, so not only does a cognize c but a also recognizes b's goal that a cognize c.

This recognition relies on agents' knowing a defeasible rule that says that

$$goal(g_1, b) \wedge cause(g_2, g_1) \supset goal(g_2, b) \tag{6}$$

That is, if an agent b has a goal g_1 and g_2 tends to cause g_1, then b may have g_2 as a goal as well.

In the case of communication, g_1 is $cognize(a, c)$ and g_2 is $present(b, x, a)$. The recipient observes the event of the presenting, uses axiom (5) to infer abductively that it is intentional, and uses axiom (6) together with schema (4) to recognize that b intends for a to cognize c.

We can get to full Gricean nonnatural meaning [6] by decomposing Rule (6) into two rules:

$$goal(g_1, b) \wedge cause(g_2, g_1) \supset goal(cause(g_2, g_1), b) \tag{7}$$

$$goal(cause(g_2, g_1), b) \supset goal(g_2, b) \tag{8}$$

That is, if an agent b has a goal g_1 and g_2 tends to cause g_1, then b may have as a goal that g_2 cause g_1. Moreover, if an agent b has as a goal that g_2 cause g_1, then b has the goal g_2.

When g_1 is $cognize(a, c)$ and g_2 is $present(b, x, a)$, a uses axioms (7) and (8) to explain the presentation; then a will recognize not only b's intention to have a cognize c but also b's intention that a do so *by virtue of* the causal relation between perceiving x and cognizing c. This is the definition of Gricean nonnatural meaning.

In order for this sort of communication to work, it must be mutually known between a and b that perceiving x causes cognizing c. Mutual belief can be characterized by three properties. First, if a group mutually believes something, each of the members believes it. Second, if a group mutually believes something, then it mutually believes that it mutually believes it; this allows one to step up to arbitrary levels of embedding of belief inside belief. Third, mutual belief can be successively approximated by shared beliefs. That is, if we both believe something, we probably but not necessarily mutually believe it; if we both believe we both believe it, it is even more likely that we mutually believe it. The more levels of embedding we add on, the more difficult it is to construct examples where mutual belief fails, and thus, the more likely it is that mutual belief holds.

Communicative conventions are causal rules having the form of (7) and (8), where g_1 is the presentation of a symbol and g_2 is the act of cognizing its meaning. Communicative conventions grow up in different social groups and become mutually believed within the groups. Thus, the structure of a communicative convention is

$$mb(s, cause(present(b, x, a), cognize(a, c))) \wedge member(a, s) \wedge member(b, s)$$

for a specific x and a specific c. That is, a social group s that agents a and b are members of mutually believes the causal relation between presenting x and cognizing c. For example, x might be a red flag with a white diagonal, s might be the community of boaters, and c the concept that there is a scuba diver below.

These communicative conventions can originate and take hold in a group in many different ways. The culture of a group consists in large part of a number of such rules.

Note that there is nothing particularly advanced about the arbitrariness of the symbol x. That is already there in the most primitive stage, in the connection between the bell and the food.

This completes the sketch of how the meaning of atomic symbols can be grounded in a theory of cognition: in our scheme, x is a symbol that means or represents c to a group of agents s. In an elaboration of Pease and Niles [22] we can express this as

$$means(x, c, s)$$

We will leave out the third argument in the development of the theory of diagrams below; the community is simply the set of people expected to be able to understand the diagrams.

Like a folk theory of belief, a folk theory of intention is a useful thing for a social animal to have. It allows individuals to predict the behavior of other individuals with some degree of accuracy. A stark example of this point is the difference in our reaction to walking down a sidewalk a few feet from a bunch of cars hurtling past at 50 miles an hour and our reaction to standing on a mountain slope a few feet from a landslide. The difference is entirely a matter of intention. Once a theory of intention has evolved, it can be deployed in the construction of a much richer theory of meaning, as described here, and for example allows agents to interpret utterances as deliberately deceptive.

We next turn to how more complex symbolic objects convey more complex meanings in different modalities.

4   Composition in Symbol Systems

An atomic symbol, i.e., one that does not have interpretable parts, corresponds to some concept. Atomic symbols can be composed in various ways, depending on the type of symbol system, and the meaning of the composition is determined by meaning of the parts and the mode of composition. These composite elements can then be components in larger structures, giving us symbolic structures of arbitrary complexity. This is illustrated in the commuting diagram of Fig. 1.

Composition in symbol systems occurs when entities x and y, meaning c_1 and c_2, respectively, are presented with a relation R_1 between them, where R_1 conveys the relation R_2 in the target domain. Thus, we have three causal relations.

$$cause(present(b, x, a), cognize(a, c_1))$$

$$cause(present(b, y, a), cognize(a, c_2))$$

$$cause(present(b, R_1(x, y), a), cognize(a, R_2(c_1, c_2)))$$

The relation $R_1(x, y)$ can be thought of as just another entity in the symbol system, so it is subject to full Gricean interpretation just as atomic symbols are, and it can similarly be involved in the conventions of some community.

With respect to the concepts invoked, we will confine ourselves here to propositional concepts. The advantage of having a flat notation in which anything can be reified is that when composite concepts are constructed, we can view this as simply

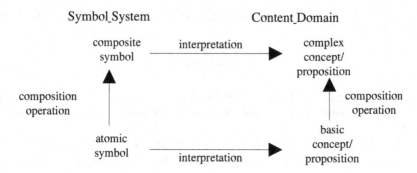

Fig. 1 Composition in symbol systems

a conjunction of what is already cognized with the new relations conveyed by the composition.

Speech (and text as spoken) takes place in time, so the only compositional relation possible on the left side of the diagram is concatenation. In discourse beyond the sentence, concatenation generally conveys a coherence relation based on causality, similarity, interlocking change of state, or figure-ground [8]. In a sense, the adjacency of the segments of discourse says that the states or events described in each segment are related somehow, and the hearer has to figure out how. The particular coherence relations that occur are the most common kinds of relations that typically obtain among states and/or events.

Within sentences, the composition of smaller units into larger units conveys primarily a predicate-argument relation between the meanings of the components. Thus, when we concatenate "men" and "work" into "men work," we are indicating that the referent of "men" is an argument or role-filler in the event denoted by "work." This view of syntax as conveying predicate-argument (and modification) relations through adjacency of constituents is elaborated in [10], in which an extensive grammar of English is developed in a manner similar to the Head-driven Phrase Structure Grammar of Pollard and Sag [24]. Section 6 describes an implementation of this idea for generating logical forms for natural language sentences.

Protolanguage [2] is more like discourse. Two words, phrases, or larger segments of natural language are adjacent, and it is up to the hearer to figure out the relation between them. Children's language in the two-word phase is an example of this; "mommy sock" might mean that this sock belongs to mommy, or that mommy should put this sock on or take it off the child, or any number of other relations. The language of panic is another example of protolanguage, for example, when someone runs out of his office shouting, "Help! Heart attack! John! My office! CPR! Just sitting there! 911! Help! Floor!" we understand the parts and hypothesize the relations that build them into a coherent scenario. The nominal compound can be seen as a relic of protolanguage in modern English; the relation between the two nouns might be a predicate-argument relation, as in "language origin", but it might be any other relation as well, as in "turpentine jar".

Hobbs [12] tells an incremental story of the evolution of syntax that begins with protolanguage. Then constraints are added to the effect that adjacency is interpreted as predicate-argument relations. Further steps extend this to discontinuous elements, as in "Pat is likely to go," where "Pat" is the subject of "go." Long-distance dependencies are seen as language-specific constraints on possible predicate-argument relations between noun phrases and adjacent clauses.

Tables are another kind of complex symbolic object. In tables, the elements in individual cells refer to some concept. The manner of composition is placement of these cells in a vertical and horizontal arrangement with other cells. Generally, the aggregate represents a set of relations: The item in a cell that is not the first in its row stands in some relation to the first element in the row. The relation is the same for all elements in that column and is often explicitly labeled by a header at the top of the column. For example, in a table of United States presidents, we might have the year 1732 in one cell. The label on the row may be "George Washington" and the label on the column "Birth date." This spatial arrangement then conveys the relation $birthdate(GW, 1732)$.

A beep in your car can mean several things—you haven't fastened your seat belt, the door is open, your lights are still on. When a sequence of beeps is combined with a property of that sequence, namely, a rapid frequency, the composite meaning can be more precise—your lights are still on.

A map is an interesting example of a complex visual symbolic object. There are underlying fields, indicated by colors, that represent regions of various political or geologic types. Icons are overlaid on these fields in a way that bears at least a topological relation to the reality represented, and perhaps a geometric relation. Generally there is a mixture of topological and geometric correspondences; the width of a road on a map is usually not proportional to its width in reality. The icons represent entities of interest. The icons can have internal structure representing different categories of entities; for example, the size of a city may be represented by the number of circles around a center dot. Labels represent names. Information is conveyed by the possible spatial relations of adjacency, distance, and orientation. For example, labels naming cities are placed near the icon representing the city.

In a process diagram, as in maps, icons represent entities of different types. Individual states may be represented by a set of icons standing in particular spatial relationships to each other, representing the relations among the entities that characterize the state. Adjacent states may be connected by arrows, the direction of the arrow indicating temporal order and thus state transitions [23, 26].

Documents are composed of blocks of text and other symbolic elements in a particular spatial relationship. Some spatial relationships are tightly constrained. Titles need to be at the top of the document; the relation conveyed is that the title somehow indicates the content of the body of the document. Paragraphs meant to be read in sequence should be adjacent (with exceptions for page and column breaks). Paragraph breaks should also correlate with the coherence structure of the text, although often in a complex way. Other spatial relationships can be looser. A diagram, map, photograph, or sidebar only needs to be relatively near the block of text describing roughly the same thing, and the ordering of such figures should respect the order of references to them in the body of the text.

Web sites and PowerPoint presentations are complex symbolic objects amenable to similar analysis.

The performance of an individual in face-to-face conversation [1] is also a complex symbolic object with its own compositional principles. The principal component symbolic elements are the speech stream, prosody, facial expressions, gaze direction, body posture, and gestures with the hands and arms. The primary mode of composition is temporal synchrony. The relationship conveyed by temporal synchrony could either be that the content of the two presentations are the same, as when someone makes a rolling gesture while describing Tweety rolling down the hill as in McNeil's experiments [20], or parallel actions, as when someone freezes her arm in midgesture to hold her turn while beginning a new clause. Lascarides and Stone [17] have shown that the possible relations among gestures and between gestures and speech are the same as the coherence relations that obtain between adjacent segments of discourse, namely, those based on causality, similarity and contrast, elaboration, interlocking change of state, and figure-ground.

A play is a more complex composite symbolic object, but its structure is equally amenable to a compositional account in terms of adjacency relations among the symbolic components and conjoined relations in the meaning.

In Sect. 6 we describe a program implementing these ideas to translate natural language into logical form. In Sect. 7 we develop more fully a theory of diagrams to illustrate the application of this theory of information structure to a special case of substantial importance. But first some comments about coreference.

5 Coreference

In a complex symbolic object, different component presentations may convey the same concept, or more generally, closely related concepts. Thus there is a coreference relation between them. The most familiar example is between two noun phrases in speech or text. On a map, the dot with a circle around it representing a city and the label with the name of the city are in a sense coreferential, both representing the city. In a process diagram, the identity of shape or color of icons in two state depictions may indicate coreference relations; they are the same entity in different stages of the process. In a document a region of a photograph, a phrase in the caption on the photograph, and a portion of the accompanying text may all be coreferential. An iconic gesture and the accompanying description in a face-to-face conversation may also be coreferential.

It is clear that the coding of coreference relations in Web sites would be very useful for retrieval. For example, we would be able to move beyond the rather haphazard results we now get from image searches.

6   Compositional Semantics

Hobbs [7,9] defines a logical formalism for representing natural language in which the logical form of a sentence is a flat conjunction of atomic propositions in which all variables are existentially quantified with the widest possible scope. What in more standard notations are represented by scope are represented in this notation by predicate-argument relations, through the liberal use of reification of such things as possible eventualities and type elements of sets.

In such a notation, it is very straightforward to see how compositional semantics can be organized around the idea that concatenation within sentences means predicate-argument relations. Morphemes provide one or more predications, that is, an n-ary predicate applied to n arguments, where the arguments are existentially quantified variables. When two adjacent strings are recognized as part of the same constituent, we have recognized the identity of two of the variables. This effects a recognition of the predicate-argument relation. Consider a simple example.

Tall men succeed.

We will ignore the plural and present tense, although they are not difficult to handle. The word "tall" contributes the proposition $tall(x_1)$. The morpheme "man" contributes $man(x_2)$. The word "succeed" contributes $succeed(x_3)$.

$$tall(x_1) \wedge man(x_2) \wedge succeed(x_3)$$

When we recognize "tall men" as an NP, this tells us that $x_1 = x_2$. When we recognize "Tall men succeed" as a clause, this tells us that $x_2 = x_3$. With a flat logical form, the function application of lambda expressions that is the staple of most treatments of compositional semantics, from Montague [21] on, in fact does no more than identify variables. Hobbs [10] presents a sizeable portion of English syntax formalized along these lines.

Nishit Rathod [25] developed a tool called LFToolkit for generating logical form from parse trees based on this idea. A very simple example of an LFToolkit rule would be something like the following:

```
<S,e-1,x-1,y-1> -> NP:<NP,x-1> VP:<VP,e-1,x-1,y-1>
```

This says that when you concatenate an NP and a VP to get a clause S, then the entity x-1 referred to by the NP is the same as the subject that the VP needs. Other rules output logical form fragments at the lexical level.

We have employed LFToolkit in a system for generating flat logical forms for natural language text, handling a substantial proportion of English syntax.

We exploit recent advances in statistical parsing by using as our input the parse trees produced by a statistical parser such as the Charniak parser [3]. However, these parsers are trained on the Penn TreeBank [18], which lacks structure crucial for generating logical form. In particular, there is no structure in the left modifiers of the head noun in an NP. No complement-adjunct distinction is made. Gap-filler relations are unmarked in long-distance dependency constructions. The label SBAR covers questions, relative clauses, subordinate clauses, and "that" clauses, among others, and all of these are realized in different ways in logical form. Finally, there are thousands of patterns because the parse trees are not binary. For example, a transitive verb phrase can have any number of adverbials after the complement.

$$VP \rightarrow \quad Verb \ NP \ ADVP \ ADVP \ ADVP \ ADVP \ ADVP \ldots \qquad (9)$$

Hence, we would have to write an unlimited number of LFToolkit rules to accommodate all of these supposed constructions.

Consequently, we first preprocess the parse tree to put it into an amenable form that contains all the right information in the places where it is needed. We insert lexical information, e.g., that a verb is transitive, into the lexical nodes, using a very large lexicon that is based on the lexicon that was used in SRI's DIALOGIC project in the 1980s but was augmented from various other resources. Lexical information is passed up the tree to the nodes where it is needed, so that, for example, the VP knows its head verb is transitive. Structural information is passed down the tree to the nodes where it is needed, for example, about what kind of SBAR a clause or verb phrase is embedded in. In addition, the tree is binarized to eliminate structures like (9).

At this point it is possible to capture a significant portion of English syntax with 200–300 LFToolkit rules.

This system has been used in demo systems and in computational linguistics classes on arbitrary English written texts in a number of domains.

When proposing a general theory of information structure, one huge challenge is to show that it can handle the complex meanings of our most elaborated symbol system, natural language. We have done just that by means of the implementation of a system for generating logical form from natural language text, based on the idea that concatenation in sentences conveys predicate-argument relations, or equivalently, the identity of locally posited, existentially quantified variables.

Another challenge for such a theory is whether the meanings of diagrams can be captured. This is the topic of the next section.

7   A Theory of Diagrams

7.1   Background Theories

In this section we develop a theory of diagrams and apply it to defining Gantt charts. A Gantt chart is a particularly common type of diagram in certain corporate cultures; it graphs tasks and subtasks in a project against a time line. We need to rely on concepts from several background theories, not all of which have been described in published papers.

Set Theory: We need one relation and one function:

$member(x, s)$: x is a member of the set s.

$card(s) = n$:   The cardinality of set s is n.

Composite Entities: A composite entity x is something that has a set of components and a set of relations among the components. We will need two relations:

$componentOf(x, s)$: x is a component of s.

$relationOf(r, s)$: r is a relation of s.

This depends on reifying relations (cf. [7]).

Scales: One-dimensional diagram objects, intervals of time, and, by extension, events are all scales and have beginnings and ends. It will be convenient to have these concepts in both relational and functional form:

$begins(x, s)$: x is the beginning of s.

$ends(x, s)$: x is the end of s.

$beginningOf(s) = x$: x is the beginning of s.

$endOf(s) = x$: x is the end of s.

Strings: We assume there are strings of characters. They are usually symbolic objects, so they can be the first argument of the predicate *means*.

Time: In the development on Gantt charts, reference is made to concepts in OWL-Time [13]. This ontology posits temporal entities (i.e., intervals and instants), the beginnings and ends of intervals, a *before* relation, Allen interval relations like *intMeets*, and temporal aggregates, which are sets of nonoverlapping, ordered temporal entities. It also handles durations and clock and calendar terms. The three predicates we need here are the following:

$TimeLine(t)$: t is the infinite interval containing all

                   temporal entities.

$atTime(e, t)$:   Event e occurs at instant t.

$calInt(t)$: t is a calendar interval, i.e., a calendar day, week,

                   month, or year.

In addition, we will need one predicate relating strings to times:

$dateStringFor(s, t)$: s is a string describing temporal entity t.

The one predicate we need from a theory of causality or processes [11] is *enables*:

$enables(e_1, e_2)$:   Event or condition e_1 enables, or is a prerequisite for,

event or condition e_2.

For Gantt charts we need a simple ontology of projects, with the following three predicates.

$Project(p)$: p is a project.

$taskIn(z, p)$: z is one of the tasks of project p.

$milestoneIn(m, p)$: m is one of the milestones of project p.

A project is a composite entity among whose components are tasks and milestones, which are events. The project and its parts can have names.

$name(s, z)$:   The string s is the name of z.

Space: The actual drawing of a diagram will involve mapping the ontology of diagrams to an ontology of space. Some portion of space will have to be chosen as the ground. This will define the vertical and horizontal directions and the *above* and *rightOf* relations. In addition, the articulation between the theory of diagrams and the theory of space would have to specify the kinds of spatial regions that realize different kinds of diagram objects.

7.2   Diagram Objects

A diagram consists of various diagram objects placed against a ground, where each diagram object has a meaning. We can take the ground to be a planar surface, which thus has points. Diagram objects can have labels placed near them, and generally they indicate something about the meaning. Diagram objects, points, frameworks, meanings, and labels are discussed in turn, and then it is shown how these can be used to define Gantt charts.

Diagram objects can be classified in terms of their dimensionality. In a spatial ontology in general, we would have to specify both a dimension of the object and the dimension of the embedding space, but in this theory of diagrams, we will take our embedding space to be a two-dimensional plane. Thus, there are three types of diagram objects:

$$0DObject(x) \supset DiagramObject(x)$$

$$1DObject(x) \supset DiagramObject(x)$$

$$2DObject(x) \supset DiagramObject(x)$$

Zero-dimensional diagram objects in diagrams are the class of diagram objects that are treated as having zero dimensions in the context of the diagram. Of course, in a spatial ontology they would actually be small regions generally with some symmetries. Three types of zero-dimensional diagram objects are dots, tickmarks, and diamonds.

$$Dot(x) \supset 0DObject(x)$$

$$Tickmark(x) \supset 0DObject(x)$$

$$Diamond(x) \supset 0DObject(x)$$

One-dimensional diagram objects in diagrams include curves.

$$Curve(x) \supset 1DObject$$

Three important kinds of curves are lines, rays (half-lines), and line segments.

$$Line(x) \supset Curve(x)$$

$$Ray(x) \supset Curve(x)$$

$$LineSegment(x) \supset Curve(x)$$

A line has no beginning or end. A ray has a unique beginning but no end. A line segment has both a unique beginning and a unique end.

$$Line(x) \supset [\neg(\exists p_1)[begins(p_1, x)] \wedge \neg(\exists p_2)[ends(p_2, x)]]$$

$$Ray(x) \supset [[(\exists ! p_1)[begins(p_1, x)] \wedge \neg(\exists p_2)[ends(p_2, x)]]]$$

$$LineSegment(x) \supset (\exists ! p_1, p_2)[begins(p_1, x)] \wedge ends(p_2, x)]$$

Beginnings and ends are points, in the sense described below.

Occasionally below it will be convenient to have functions for line segments corresponding to the *begins* and *ends* relations.

$$LineSegment(x) \supset (\forall p)[beginningOf(x) = p \equiv begins(p, x)]$$

$$LineSegment(x) \supset (\forall p)[endOf(x) = p \equiv ends(p, x)]$$

It will be useful to have a term *Linear* that covers all three types of linear diagram objects.

$$Linear(x) \equiv [Line(x) \vee Ray(x) \vee LineSegment(x)]$$

A line segment "in" a linear diagram object is one that is wholly contained in it.

lineSegmentIn(x, y)

$$\equiv LineSegment(x) \wedge Linear(y) \wedge (\forall p)[pointIn(p, x) \supset pointIn(p, y)]$$

Another kind of curve is an arrow.

$$Arrow(x) \supset Curve(x)$$

An arrow has a specific unique beginning and end.

$$Arrow(x) \supset (\exists ! p_1, p_2)[begins(p_1, x) \wedge ends(p_2, x)]$$

An arrow may be straight or curved. It may overlap in part with other arrows. An arrow is similar to a line segment but is realized in the underlying spatial ontology differently, e.g., with a small triangle for its end.

It will be useful to talk about an arrow as a ternary relation.

$$arrow(x, p_1, p_2) \equiv [Arrow(x) \wedge begins(p_1, x) \wedge ends(p_2, x)]$$

Diagrams are composite entities whose components are diagram objects.

$$Diagram(d) \wedge componentOf(x, d) \supset DiagramObject(x)$$

7.3   Points and the *at* Relation

A ground consists of *points* and any diagram object consists of points, in some loose sense of "consist of"; that is, for any ground and any diagram object there is a corresponding set of points.

$$[Ground(x) \vee DiagramObject(x)] \supset (\exists s)(\forall p)[member(p, s) \supset pointIn(p, x)]$$

A zero-dimensional object has exactly one point in it.

$$0DObject(x) \supset (\exists ! p)pointIn(p, x)$$

For convenience we will say that the single point in a zero-dimensional object both begins and ends it.

$$0DObject(x) \supset (\forall p)[pointIn(p, x) \equiv [begins(p, x) \wedge ends(p, x)]]$$

Points are not diagram objects.

The beginnings and ends of linear objects are points.

$$begins(p, x) \land Linear(x) \supset pointIn(p, x)$$
$$ends(p, x) \land Linear(x) \supset pointIn(p, x)$$

Points in the ground are partially ordered by an *above* relation and a *rightOf* relation.

$$above(p_1, p_2, g) \supset Ground(g) \land pointIn(p_1, g) \land pointIn(p_2, g)$$
$$rightOf(p_1, p_2, g) \supset Ground(g) \land pointIn(p_1, g) \land pointIn(p_2, g)$$

A linear object is horizontal if no point in it is above any other. Similarly, vertical.

$$horizontal(x, g) \equiv Linear(x)$$
$$\land \neg(\exists p_1, p_2)[pointIn(p_1, x) \land pointIn(p_2, x) \land above(p_1, p_2, g)]$$
$$vertical(x, g) \equiv Linear(x)$$
$$\land \neg(\exists p_1, p_2)[pointIn(p_1, x) \land pointIn(p_2, x) \land rightOf(p_1, p_2, g)]$$

A horizontal ray all of whose points are to the right of its beginning is a rightward positive ray.

$$[ray(x) \land horizontal(x, g) \land begins(p_0, x)$$
$$\land(\forall p)[pointIn(p, x) \supset [p = p_0 \lor rightOf(p, p_0, g)]]$$
$$\supset rtPositive(x, g)$$

A vertical ray all of whose points are above its beginning is a upwardly positive ray. A vertical ray all of whose points are below its beginning is a downwardly positive ray.

$$[ray(x) \land vertical(x, g) \land begins(p_0, x)$$
$$\land(\forall p)[pointIn(p, x) \supset [p = p_0 \lor above(p, p_0, g)]]$$
$$\supset upPositive(x, g)$$
$$[ray(x) \land vertical(x, g) \land begins(p_0, x)$$
$$\land(\forall p)[pointIn(p, x) \supset [p = p_0 \lor above(p_0, p, g)]]$$
$$\supset dnPositive(x, g)$$
$$rtPositive(x, g) \supset ray(x)$$
$$upPositive(x, g) \supset ray(x)$$
$$dnPositive(x, g) \supset ray(x)$$

A special kind of line segment needed for Gantt charts is a horizontal bar.

$$HBar(x) \supset (\exists g)[horizontal(x, g) \wedge LineSegment(x)]$$

When realized spatially, it will generally be thicker than other line segments.

Diagrams are constructed by placing points in diagram objects at points in the ground or in another diagram object. The *at* relation expresses this.

$$at(p_1, p_2)$$

$$\supset (\exists x_1, x_2)[pointIn(p_1, x_1) \wedge pointIn(p_2, x_2) \wedge DiagramObject(x_1)$$

$$\wedge[Ground(x_2) \vee DiagramObject(x_2)] \wedge x_1 \neq x_2]$$

The relation *at* can be extended to zero-dimensional objects in the obvious way.

$$0DObject(x) \supset (\forall p)[at(x, p) \equiv (\exists p_1)[pointIn(p_1, x) \wedge at(p_1, p)]]$$

Typically, frameworks (see below) will be placed with respect to some points in the ground, and other diagram objects will be placed with respect to the framework or other diagram objects.

The relations of a diagram as a composite entity include its *at* relations. To say this formally we can reify the *at* relation. Thus, $at'(r, p_1, p_2)$ means that r is the *at* relation between p_1 and p_2. We can then say that r is a member of the relations of the diagram.

$$at'(r, p_1, p_2) \wedge relationOf(r, d)$$

7.4 Frameworks

Many diagrams have an underlying framework with respect to which diagram objects are then located, e.g., the lat-long framework on maps. A framework is a set of objects in a particular relationship to each other.

$$Framework(f)$$

$$\supset (\exists s)(\forall x)[member(x, s) \supset DiagramObject(x) \wedge componentOf(x, f)]$$

One very important kind of framework is a coordinate system. Here I will characterize only a rectilinear cooordinate system.

$$RCoordinateSystem(f) \supset Framework(f)$$

$$RCoordinateSystem(f) \supset (\exists g)[Ground(g) \land groundFor(g, f)]$$

$$RCoordinateSystem(f) \land groundFor(g, f)$$
$$\supset (\exists x, y)[xAxisOf(x, f) \land yAxisOf(y, f) \land rtPositive(x, g)$$
$$\land[upPositive(y, g) \lor dnPositive(y, g)]]$$

$$xAxisOf(x, f) \supset RCoordinateSystem(f) \land componentOf(x, f)$$

$$yAxisOf(y, f) \supset RCoordinateSystem(f) \land componentOf(y, f)$$

An x-axis and a y-axis are both axes.

$$xAxisOf(x, f) \supset axis(x)$$

$$yAxisOf(y, f) \supset axis(y)$$

Two points have the same x-coordinate if there is a vertical line that contains both of them. Similarly, same y-coordinate.

$$sameX(p_1, p_2, f)$$
$$\equiv (\exists l, g)[groundFor(g, f) \land vertical(l, g) \land pointIn(p_1, l) \land pointIn(p_2, l)]$$

$$sameY(p_1, p_2, f)$$
$$\equiv (\exists l, g)[groundFor(g, f) \land horizontal(l, g) \land pointIn(p_1, l) \land pointIn(p_2, l)]$$

The xValue of a point p is a point p_1 in the x axis with the same x-coordinate. Similarly for the yValue.

$$xValue(p_1, p, f) \equiv (\exists x)[sameX(p_1, p, f) \land pointIn(p_1, x) \land xAxisOf(x, f)]$$

$$yValue(p_2, p, f) \equiv (\exists y)[sameY(p_2, p, f) \land pointIn(p_2, y) \land yAxisOf(y, f)]$$

It will be convenient below to talk about the yValue of a horizontal line segment.

$$LineSegment(h) \land horizontal(h, g) \land groundFor(g, f)$$
$$\supset (\forall p_2)[yValue(p_2, h, f) \equiv (\exists p)[pointIn(p, h) \land yValue(p_2, p, f)]]$$

7.5 Meanings

Associated with every object in a diagram is its meaning. Meaning for diagrams is thus a function mapping diagram objects into entities provided by some other ontology. Meaning is conveyed by the predication $means(x, c)$ introduced above, where x is a diagram object. There are no constraints on the second argument of *means*; it just has to be an entity in some ontology.

$$DiagramObject(x) \supset (\exists c)means(x, c)$$

The meanings of the *at* relations in a diagram will be specified by means of axioms having the following form:

$$at(x, y) \wedge p(x) \wedge q(y) \wedge means(x, a) \wedge means(y, b) \supset r(a, b)$$

That is, if a p-type diagram object x is at a q-type diagram object y in a diagram, then if x means a and y means b, then there is an r relation between a and b.

Axes in a coordinate system generally mean some set in another ontology. That set may be unordered (a set of tasks), discrete and linearly ordered (months), or continuous (time).

7.6   Labels

A label is a textual object that can be associated with objects in a diagram. The two basic facts about labels cannot be defined with precision without making reference to the cognition of the reader of the diagram.

1. A label is placed near the object it labels, in a way that allows the reader of the diagram to uniquely identify that object.
2. The content of the label as a string bears some relation to the meaning of the object that it labels, in that perceiving the string causes one to think of the meaning.

Specifying the first of these completely is a very hard technical problem [4]. For example, often on a map one cannot correctly associate the name of a town with a dot on the map without doing the same for all nearby towns, and associating a curve on a map with the name of a road often requires abductive inferences about shortest paths and consistency of line thickness. Here we will simply say that a label can be placed *at* an object and leave it to component-specific computation to determine what *at* means in some context.

$$label(l, x) \supset string(l) \wedge DiagramObject(x) \wedge at(l, x)$$

The second property of labels is also a difficult technical, or even artistic, problem. But a very common subcase is where the label is a name. The whole purpose of a name is to cause one to think of the object when one perceives the name, so it serves well for this property of labels.

$$label(l, x) \supset (\exists c)[means(l, c) \wedge means(x, c)]$$

7.7   Gantt Charts

A Gantt chart g for a project p is a diagram that consists of several types of components. It has a rectilinear coordinate system f where the x-axis is rightward positive and the y-axis is upward or downward positive. (The x-axis can appear at the top or the bottom of the chart.) The meaning of the x-axis is the time line or some other periodic temporal aggregate, and the meaning of the y-axis is a set of tasks in the project.

$GanttChart(g, p)$

$\supset Diagram(p) \wedge Project(p)$

$\quad \wedge (\exists f, x, y, t, s)[RCoordinateSystem(f) \wedge componentOf(f, g) \wedge xAxisOf(x, f)$

$\quad\quad \wedge rtPositive(x) \wedge means(x, t) \wedge TimeLine(t) \wedge yAxisOf(y, f)$

$\quad\quad \wedge [upPositive(y) \vee dnPositive(y)] \wedge means(y, s)$

$\quad\quad \wedge (\forall z)[member(z, s) \supset taskIn(z, p)]]$

A Gantt chart has horizontal bars representing the interval during which a task is executed.

$GanttChart(g, p) \wedge RCoordinateSystem(f) \wedge componentOf(f, g)$

$\quad \supset (\exists s)(\forall b)[member(b, s) \supset componentOf(b, g) \wedge HBar(b)$

$\quad\quad \wedge (\exists r_1, z, p_1, t_1, q_2, t_2)[yValue(r_1, b, f) \wedge means(r_1, z) \wedge taskIn(z, p)$

$\quad\quad \wedge xValue(p_1, beginningOf(b), f) \wedge means(p_1, t_1) \wedge begins(t_1, z)$

$\quad\quad \wedge xValue(q_1, endOf(b), f) \wedge means(q_1, t_2) \wedge ends(t_2, z)]]$

Because a task is an event, OWL-Time allows instants as the beginnings and ends of tasks. This axiom says that a Gantt chart has a set of components which are horizontal bars representing tasks and the beginning of the bar represents the starting time of the task and the end of the bar represents the finishing time of the task.

Similarly, a Gantt chart has diamonds representing milestones.

$GanttChart(g, p) \wedge RCoordinateSystem(f) \wedge componentOf(f, g)$

$\quad \supset (\exists s)(\forall d)[member(d, s) \supset componentOf(d, g) \wedge Diamond(d)$

$\quad\quad \wedge (\exists m, r_1, r_2, t_1)[yValue(r_2, d, f) \wedge means(r_2, m) \wedge milestone(m, p)$

$\quad\quad \wedge xValue(r_1, d, f) \wedge means(r_1, t_1) \wedge atTime(m, t_1)]]$

We can call bars and diamonds "task icons."

$$taskIcon(x) \equiv [HBar(x) \vee Diamond(x)]$$

A Gantt chart often has arrows going from the end of one bar to the beginning of another, indicating the first bar's task is a prerequisite for the second bar's task. A diamond can also be the source and/or target of an arrow.

$$GanttChart(g, p) \wedge RCoordinateSystem(f) \wedge componentOf(f, g)$$
$$\supset (\exists s)(\forall a)[member(a, s) \supset componentOf(a, g)$$
$$\wedge(\exists x, z_1, p_1, y, z_2, p_2)[arrow(a, p_1, p_2) \wedge taskIcon(x) \wedge componentOf(x, g)$$
$$\wedge means(x, z_1) \wedge ends(p_1, x) \wedge taskIcon(y) \wedge componentOf(y, g)$$
$$\wedge means(y, z_2) \wedge begins(p_2, y) \wedge enables(z_1, z_2)]]$$

A bar in a Gantt chart may have labels for the date at its beginning and end.

$$GanttChart(g, p) \wedge HBar(b) \wedge ComponentOf(b, g)$$
$$\supset [(\exists s_1)(\forall l_1)[member(l_1, s_1) \equiv (\exists p_1, q_1, t_1)[begins(p_1, b) \wedge label(l_1, p_1)$$
$$\wedge xValue(q_1, p_1) \wedge means(q_1, t_1) \wedge dateStringFor(l_1, t_1)]$$
$$\wedge card(s_1) < 2]$$
$$\wedge(\exists s_2)(\forall l_2)[member(l_2, s_2) \equiv (\exists p_2, q_2, t_2)[ends(p_2, b) \wedge label(l_2, p_2)$$
$$\wedge xValue(q_2, p_2) \wedge means(q_2, t_2) \wedge dateStringFor(l_2, t_2)]$$
$$\wedge card(s_2) < 2]]$$

The cardinality statement is a way of saying there is either zero or one label. Similarly, a diamond in a Gantt chart may have a label for a date.

$$GanttChart(g, p) \wedge Diamond(d) \wedge componentOf(d, g)$$
$$\supset (\exists s)[(\forall l)[member(l, s) \equiv (\exists q, t)[label(l, d) \wedge xvalue(q, d) \wedge means(q, t)$$
$$\wedge dateStringFor(l, t)]]$$
$$\wedge card(s) < 2]$$

Points on the y-axis of a Gantt chart can be labeled with task names.

$$[GanttChart(g, p) \wedge RCoordinateSystem(f) \wedge componentOf(f, g) \wedge yAxisOf(y, f)$$
$$\wedge pointIn(p_1, y)]$$
$$\supset (\exists s)[(\forall l)[member(l, s) \equiv (\exists z)[means(p_1, z) \wedge name(l, z)]]$$
$$\wedge card(s) < 2]$$

Points in the x-axis of a Gantt chart can be labeled with dates or times.

$[GanttChart(g, p) \land RCoordinateSystem(f) \land componentOf(f, g) \land xAxisOf(x, f)$

$\qquad \land pointIn(p_1, x)]$

$\supset (\exists s)[(\forall l)[member(l, s) \equiv (\exists t)[label(l, p_1) \land means(p_1, t)$

$\qquad \land dateStringFor(l, t)]]$

$\qquad \land card(s) < 2]$

Line segments in the x-axis of a Gantt chart can be labeled with calendar intervals.

$[GanttChart(g, p) \land RCoordinateSystem(f) \land componentOf(f, g) \land xAxisOf(x, f)$

$\qquad \land lineSegmentIn(s_1, x)]$

$\supset (\exists s)[(\forall l)[member(l, s) \equiv (\exists t)[means(s_1, t) \land label(l, s_1)$

$\qquad \land calInt(t) \land dateStringFor(l, t)]]$

$\qquad \land card(s) < 2]$

Further elaborations are possible. The labels can have internal structure. For example, labels for subtasks may be indented. Labels for time intervals may be broken into a line for months, a line below for weeks, and so on.

8 Modalities, Media, and Manifestations

In order for communication to work, perception of the symbol must occur. Humans are able to perceive optical, acoustic, and chemical phenomena, as well as pressure and temperature. Of these modalities the optical and acoustic are by far the most important, because they offer the richest possibilities for composition. Artifact agents of course have other modalities.

Communication requires one or more devices. There must be a manner in which the presentation is carried out. Allwood [1] categorizes these into primary, secondary, and tertiary. The primary devices or media are the ones that are human body parts and processes. The voice is used for grunts, speech and song. The hands, arms, body, face, and head are used for gesture. Even at this level some encoding must be done; we need to find words for the concepts we wish to convey, and these must be mapped into sequences of articulatory gestures.

The secondary media are those involving devices external to the human body, such as marks on paper as in writing and drawings, computer terminals, telephones, videotape, and so on. These typically require multiple encodings, where the final code is known to the intended audience. The various media have different advantages and disadvantages that can be exploited for different kinds of represented content. For example, visual spatial representations can exploit more dimensions for conveying relationships than can auditory temporal representations. Hovy and Arens [16] catalog many of these correlations.

Allwood also mentions tertiary media, including paintings, sculptures, and aesthetic designs of artifacts such as chairs. These are probably just secondary media where the content that is represented is much more difficult to capture in words.

We have a strong tendency to group together classes of symbolic entities that share the same property, especially their content, and think of the aggregates as individuals in their own right. It is probably better in an ontology of symbolic entities to view these as first-class individuals that themselves represent a particular content. Other symbolic entities may be manifestations of these individuals. The predicate *manifest* is a transitive relation whose principal property is that it preserves content.

$$manifest(x_1, x) \land means(x, c, s) \supset means(x_1, c, s)$$

(This does not take into account translations, where the s's differ.)

Thus, to use the example of Pease and Niles [22], there is an entity called *Hamlet*. The performance of *Hamlet* manifests *Hamlet*. The performance of *Hamlet* in a particular season by a particular company manifests that, and a performance on a particular night may manifest that. A videotape of that particular performance manifests the performance, and every copy of that videotape manifests the videotape. A similar story can be told about the script of the play, a particular edition of the script, and a particular physical book with that edition of the script as its content.

The above rule should be thought of as defeasible, because variations exist, lines can be dropped, and printer's errors occur. More precisely, if some proposition occurs in the content of one symbolic entity then defeasibly it occurs in the content of symbolic entities that manifest it.

9 Summary

Information structure is one of the most basic domains in an ontology of the everyday world, along with such domains as space and time. It should be anchored in an ontology of commonsense psychology, as we have tried to sketch here, and there should be an account of how complex symbolic entities can be composed out of simpler symbolic entities in various modalities and combinations of modalities. We have demonstrated the utility of this framework with respect to two of the most complex symbolic systems, natural language, and diagrams. Because Web sites are an especially complex sort of symbolic entity, we can expect this kind of theory to be very significant in the development of the modes of access to Web resources.

Acknowledgments We have profited from discussions with and work by Eduard Hovy, Nishit Rathod, Midhun Ittychariah, and Paul Martin. The opinions expressed here, however, are entirely our own. This material is based in part upon work supported by the Defense Advanced Research Projects Agency (DARPA), through the Department of the Interior, NBC, Acquisition Services Division, under Contract No. NBCHD030010. It was also supported in part by DARPA's "Learning by Reading: The Möbius Project" (NBCHD030010 TO #0007), and in part under the IARPA (DTO) AQUAINT program, contract N61339-06-C-0160.

References

1. Allwood, J.: Bodily communication–dimensions of expression and content. In: Grandström, B., House, D., Karlsson, I. (eds.) Multimodality in Language and Speech Systems. Kluwer, Dordrecht (2002)
2. Bickerton, D.: Language and Species. University of Chicago Press, IL (1990)
3. Charniak, E.: Statistical parsing with a context-free grammar and word statistics. In: Proceedings, AAAI-97, Fourteenth National Conference on Artificial Intelligence, pp. 598–603. AAAI Press/MIT Press, Menlo Park (1997)
4. Edmondson, S., Christensen, J., Marks, J., Shieber, S.M.: A general cartographic labeling algorithm. Cartographica 33(4), 13–23 (1997)
5. Gordon, A., Hobbs, J.R.: Formalizations of commonsense psychology. AI Mag. 25, 49–62 (2004)
6. Grice, P.: Meaning. In: Studies in the Way of Words. Harvard University Press, Cambridge (1989)
7. Hobbs, J.R.: Ontological promiscuity. In: Proceedings, 23rd Annual Meeting of the Association for Computational Linguistics, pp. 61–69. Chicago, Illinois (1985a)
8. Hobbs, J.R.: On the coherence and structure of discourse. Report No. CSLI-85-37, Center for the Study of Language and Information, Stanford University (1985b)
9. Hobbs, J.R.: The logical notation: ontological promiscuity. Available at http://www.isi.edu/~hobbs/disinf-tc.html (2003)
10. Hobbs, J.R.: The syntax of English in an abductive framework. Available at http://www.isi.edu/~hobbs/disinf-chap4.pdf (2005a)
11. Hobbs, J.R.: Toward a useful concept of causality for lexical semantics. J. Semant. 22, 181–209 (2005b)
12. Hobbs, J.R.: The origin and evolution of language: a plausible, strong AI account. In: Arbib, M.A. (ed.) Action to Language via the Mirror Neuron System, pp. 48–88. Cambridge University Press, Cambridge (2006)
13. Hobbs, J.R., Pan, F.: An ontology of time for the semantic web. ACM Trans. Asian Lang. Inform. Process. 3(1), 66–85 (2004)
14. Hobbs, J.R., Stickel, M., Appelt, D., Martin, P.: Interpretation as abduction. Artif. Intell. 63(1–2), 69–142 (1993)
15. Hobbs, J.R., Croft, W., Davies, T., Edwards, D., Laws, K.: Commonsense metaphysics and lexical semantics. Comput. Linguist. 13(3–4), 241–250 (1987)
16. Hovy, E.H., Arens, Y.: When is a picture worth a thousand words?–allocation of modalities in multimedia communication. AAAI spring symposium on human-computer interactions, Palo Alto, CA (1990)
17. Lascarides, A., Stone, M.: Formal semantics of iconic gesture. In: Schlangen, D., Fernandez, R. (eds.) Proceedings, BRANDIAL 06: 10th Workshop on the Semantics and Pragmatics of Dialogue, pp. 64–71. Potsdam (2006)
18. Marcus, M.: The Penn TreeBank project. Available at http://www.cis.upenn.edu/~treebank/ (2009)
19. McCarthy, J.: Circumscription: a form of nonmonotonic reasoning. In: Ginsberg, M. (ed.) Artificial Intelligence, vol. 13, pp. 27–39. Reprinted in Readings in Nonmonotonic Reasoning, pp. 145–152. Morgan Kaufmann, Los Altos (1980)
20. McNeil, D. (ed.): Language and Gesture. Cambridge University Press, Cambridge (2000)
21. Montague, R.: The proper treatment of quantification in ordinary English. In: Thomason, R.H. (ed.) Formal Philosophy: Selected Papers of Richard Montague, pp. 247–270. Yale University Press, New Haven (1974)
22. Pease, A., Niles, I.: Practical semiotics: a formal theory. In: Proceedings of the International Conference on Information and Knowledge Engineering (IKE '02), Las Vegas, Nevada, June 24–27 (2002)

23. Pineda, L., Garza, G.: A model for multimodal reference resolution. Comput. Linguist. **26**(2), 139–194 (2000)
24. Pollard, C., Sag, I.A.: Head-Driven Phrase Structure Grammar. University of Chicago Press and CSLI Publications, Chicago (1994)
25. Rathod, N.: LFToolkit. Available at http://www.isi.edu/~hobbs/LFToolkit/index.html (2008)
26. Wahlster, W., André, E., Finkler, W., Profitlich, H.-J., Rist, T.: Plan-based integration of natural language and graphics generation. Artif. Intell. **63**(1–2), 387–427 (1993)

Mechanics and Mental Change

Jon Doyle

In memoriam
Joseph Arthur Schatz, 1924–2007

Abstract Realistic human rationality departs from ideal theories of rationality and meaning developed in epistemology and economics because in human life deliberation takes time and effort, ignorance and inconsistency do not deter action, and learning takes time and slows with time. This paper examines some theories of mental change with an eye to assessing their adequacy for characterizing realistic limits on change and uses a simple kind of reasoning system from artificial intelligence to illustrate how mechanical concepts, including mental inertia, force, work, and constitutional elasticity, provide a new language and formal framework for analyzing and specifying limits on cognitive systems.

1 Vive la Résistance

The ideal actors on the stage of human imagination exhibit courageousness, decisiveness, integrity, generosity, an ability to think clearly and rapidly as they act, and an ability to change direction instantly should danger or opportunity warrant. We admire heroes and heroines and tell their stories partly to celebrate their attainment of these qualities, for many people exhibit courageousness, decisiveness, integrity, generosity, and clear thinking in small matters, and so appreciate the joy they imagine the hero and heroine must feel in larger matters. But an ability to change direction instantly? The size and profitability of the self-help section of almost any bookstore attests to the trouble most people have in changing their

J. Doyle (✉)
Department of Computer Science, North Carolina State University, Raleigh, NC, USA
e-mail: Jon_Doyle@ncsu.edu

B.-O. Küppers et al. (eds.), *Evolution of Semantic Systems*,
DOI 10.1007/978-3-642-34997-3_7, © Springer-Verlag Berlin Heidelberg 2013

behavior. Indeed, some find it easier to change behavior in large ways than in small, but few find change easy, either to effect or to accept.

The prominence of resistance to change in human nature raises doubts about the standard conception of ideal rationality in thought and action. The foundations of the ideal rationality put forward by decision theory and economics suppose that any bit of new information can lead to arbitrarily large changes arbitrarily quickly. One can easily tell someone "Everything you know is wrong!", but few outside of fantastic fiction can act on such advice even if they believe it. Real people can accept new information without connecting it with prior beliefs to draw new consequences or to notice inconsistencies and must search and think carefully to ensure that they do not overlook some relevant fact in making decisions.

Resistance to change therefore impacts the very nature of human belief. Inferences based on mundane prejudices and stereotypes permeate human knowledge and provide commonly useful conclusions unless one makes the effort to distinguish the atypical aspects of the current circumstances. Failure to find or acknowledge these atypical aspects thus leads to systems of belief corrupted by inappropriate assumptions. More generally, humans organize knowledge into numerous hierarchies of abstractions that reflect commonalities and differences in meanings. Failure to find or acknowledge commonalities and to restructure abstractions to reflect them produces conceptual organizations that omit useful connections between concepts and so impede effective recognition of and response to changing circumstances.

A robust theory of rational reasoning and action falls short without a good account of the origins and character of resistance to change. Philosophy and psychology have developed several ways of understanding resistance to change, especially the notions of habit, refraction, and entrenchment, but suffer a striking omission: the long-standing notions of inertia and force in common informal usage in everyday descriptions. Mathematicians from Galileo to Euler showed how to use these concepts precisely in characterizing the behavior of physical bodies, but psychology has suffered from the lack of such concepts except for use in informal metaphor and analogy. One sees this lack clearly in the case of theories of ideal rationality, in which the characteristic unbounded response to bounded impetus shows none of the proportional response embodied in mechanical laws of force and inertia.

The following presents an account of resistance to change in psychological systems that reveals an underlying mechanical character to psychology. This account employs an extension of the mathematical formulation of mechanics to cover minds and persons with mechanical minds and bodies. As detailed in [14], which develops the formalism in axiomatic terms along with the appropriate mathematical structures, the extended mechanics separates the fundamental properties and structures of position, motion, mass, and force from special characteristics of continuous space and extends these notions to hybrid systems that combine different types of positions, masses, and forces. The presentation here omits the axiomatics in favor of an annotated example of mechanics in a specific type of reasoner developed in artificial intelligence.

2 Active and Passive Resistance

We set the stage for an examination of the mechanical account of resistance to change by reviewing briefly the principal accounts of resistance developed in mental philosophy and psychology without reference to mechanical notions, namely the notions of habit and refraction.

Although conscious thought focuses on our deliberate reasoning and action, habits form a major element of both thought and action, to an extent that some have regarded most or all of thinking as occurring through the action of complex sets of habits. Hume famously saw the foundations of reasoning in terms of experiential associations between one condition and another that developed customs or habits of thinking. Behaviorist psychology later expanded on this notion to interpret all or almost all behavior as occurring through the combined operation of sets of habits, with positive or negative reinforcement modifying behavior by modifying mediating habits. Artificial intelligence adapted notions of habits by employing collections of automated antecedent-consequent rules [25] to replicate certain types of reasoning, and by formalizing behavioristic action in terms of hierarchies of routine behavior [2, 38].

Computational rule engineers viewed the problem of training more broadly than the behaviorists, typically looking first for human informants to state concepts, conditions, and rules explicitly, followed by iterative tailoring of behavior by the engineer or by means of automated learning mechanisms. Modern neurophysiology reinforces the importance of habit by identifying neurons as stimulus-response units and by observing how repetitive usage patterns shape the stimulus sensitivity, the response function, and the network of neural connections.

Development of ideas of habit across the years has yielded an appreciation of the power of individually simple and specialized habits to work together to produce complex and sometimes intelligent behavior, to the point that earlier conceptions of human intelligence as formed mainly by deliberate reasoning modulated by peripheral influences of habits have been upended into conceptions of human intelligence as primarily habitual behavior modulated by occasional elements of deliberate reasoning. Although this transforms the notion of habit into something much more complex and subtle than Hume and others might have had in mind, this deeper understanding of the power of habit has not been accompanied by a similar appreciation of the limits such complex habits place on the power of the reasoner to change thought and action.

Of course, people are not mere creatures of habit; they are stubborn too, exhibiting refractory and willful behavior of sorts lamented throughout history. Human refraction involves active resistance to imposed change that seeks to nullify the imposition and, in willful behavior, maintain or even strengthen current attitudes and activities.

Few philosophers have devoted much attention to the nature of active stubbornness or refraction, with Shand [34] a prominent exception in his discussion of reactions generated by various circumstances in persons of different mental character. Active resistance plays a greater role in artificial intelligence, notably in Minsky's [24] theory of the "society of mind," but few have sought to formalize refraction directly. The most relevant formalizations do not address reaction as much as they address conflict, specifically decision making in which mental subagencies argue with each other [4] or have conflicting preferences [15]. Although one can view refraction through these conceptual lenses, this view does not capture the relation between the imposed change and the reaction to it.

3 Formalizing Resistance to Change

Natural philosophers formalized resistance to change in a variety of familiar ways. Physicists identify the notion of inertia as a property of all matter that characterizes resistance to change of motion. Elasticity characterizes forces generated to restore deformed configurations to undeformed ones, such as the force proportional to displacement characterized by Hooke's law for springs. Friction characterizes forces generated by motion that act against the motion. These formalizations of resistance to change have been unavailable to mental philosophy and artificial intelligence, and this unavailability has led these fields to develop formal methods useful for characterizing resistance to change in phenomenological terms. The following briefly recounts these formal methods.

To simplify the discussion, we initially restrict attention to change of belief in logically conceived reasoners with perfect inferential abilities, and later widen our view to aspects of mental states other than belief and to change in nonideal reasoners. Perfect inferential abilities means that the reasoner knows all consequences of its beliefs. In such theories, one formalizes states of belief as deductively closed and consistent sets of beliefs or "theories" $A = Cn(A)$. General changes of mind can thus take one set A into a new set A' that involves adding and removing multiple statements.

A further simplification analyzes complicated changes into smaller changes that begin with changes due to adding or removing a single belief. There are three types of such smaller changes. One can *add* a statement x to A, denoting the result as $A + x$, simply by taking the deductive closure of the set A extended by x, that is, defining

$$A + x = Cn(A \cup \{x\}). \tag{1}$$

Obviously, simple addition is inappropriate if x is inconsistent with A, so the more useful and general types of change attempt to ensure a consistent result. The operation of *contraction*, denoted $A \doteq x$ removes x from A if possible, and the operation of *revision*, denoted $A \dotplus x$, consistently adds x to A by removing conflicting beliefs if necessary. The commonly intended connection between contraction, revision, and addition is given by the Levi identity

$$A \dotplus x = (A \doteq \neg x) + x, \tag{2}$$

stating that revision by x should be the same as adding x after first removing anything contradicting x.

Simply naming these types of operations does not define them, inasmuch as one might be able to remove a statement from a theory in many different ways, up to and including removal of all statements except the tautologies, which cannot ever be removed from deductively closed theories. Accordingly, the formal development of these operations begins by understanding the types of changes that fit with each of these operations.

We will divide the phenomenological approaches to analyzing mental change into two subclasses, *comparative* theories that provide means comparing the relative size of different changes, and *characteristic* theories that seek to characterize the superficial properties of the starting and ending states connected by changes. We will later contrast these phenomenological approaches with *substantial* theories that seek to obtain comparative and characteristic properties of revision from more detailed assumptions about the substance and structure of mental states.

3.1   Comparative Theories of Mental Change

The guiding intuition of the comparative theories is Quine's [29] principle of *minimum mutilation*, in which one seeks the "smallest" change that accomplishes the purpose of the operation.

A simple and early example of the comparative approach is Rescher's [32] interpretation of hypothetical and counterfactual reasoning, which assumed a weak preference preordering (reflexive and transitive relation) over all sets of beliefs. To determine what conditions hold in some hypothetical or counterfactual situation, one examines maximal subsets of current beliefs consistent with the hypothetical or counterfactual hypothesis (such as possible contractions) and tries to find subsets that are maximally preferred among these. Rescher called these preferred maximal consistent subsets. One then looks to see if combining each preferred maximal consistent subset with the hypothesis of the hypothetical yields the conclusion of the hypothetical.

Lewis [22] based his semantics for counterfactuals on the notion of *comparative similarity* relations among possible worlds. A comparative similarity relation consists of a ternary relation

$$X \preceq_W Y \qquad (3)$$

over possible worlds such that each binary relation $\preceq_W$ is a preordering of possible worlds, that is, satisfies the condition

$$X \preceq_W X \qquad (4)$$

for each world X and the condition that

$$X \preceq_W Z \tag{5}$$

whenever

$$X \preceq_W Y \preceq_W Z \tag{6}$$

for each X, Y, Z. Such a preorder provides a notion of similarity relative to each possible world if it satisfies the additional requirement that the preorder for each world be origin-minimizing, that is, assigning minimal rank to its "center" or "origin", for formally, that

$$W \preceq_W X \tag{7}$$

for each W and X.

These requirements encompass similarity obtained from a distance function or metric d by defining

$$X \preceq_W Y \tag{8}$$

to hold iff

$$d(W, X) \leq d(W, Y), \tag{9}$$

but the axioms of comparative similarity do not require any structure of distance, much less Euclidean distance, on the set of possible worlds. A comparative similarity relation provides a way to compare changes with respect to the same origin, but need not provide any way of comparing changes with different starting points. One can capture that broader range of comparisons instead by considering a preorder over all pairs of worlds, namely a binary relation $\preceq$ satisfying the conditions

1. $(X, Y) \preceq (X, Y)$,
2. $(X, Y) \preceq (X'', Y'')$ whenever $(X, Y) \preceq (X', Y') \preceq (X'', Y'')$, and
3. $(X, X) \preceq (X, Y)$ (or even the stronger condition $(X, X) \preceq (Y, Z)$).

One easily verifies that each such a preorder over pairs of worlds induces a comparative similarity relation by defining

$$Y \preceq_X Z \tag{10}$$

to hold just in case

$$(X, Y) \preceq (X, Z). \tag{11}$$

Comparative similarity relations, whether based at worlds or over pairs of worlds, all admit numerical representations in the usual way. We say a real-valued binary function s represents a point-based similarity relation $\preceq$ just in case we have that

$$s(X, Y) \le s(X, Z) \quad \text{whenever} \quad Y \preceq_X Z . \tag{12}$$

We say that s represents a pair-based similarity relation $\preceq$ just in case we have that

$$s(X, Y) \le s(Z, W) \quad \text{whenever} \quad (X, Y) \preceq (Z, W) . \tag{13}$$

Obviously such representations are purely ordinal in character and can make comparable alternatives that were incomparable in the represented partial preorder. We may thus regard the formalism of pair-based comparative similarity as a simple (even simple-minded) means with which to describe different levels of difficulty when moving from one world to another.

Although the notion of similarity of worlds does not in itself rest on any notion of difficulty or of resistance to change, at least some notions of similarity and dissimilarity track the ease and difficulty of changing from one mental state to another, as will be seen later.

3.2   Characteristic Theories of Mental Change

To go further than the abstract notion of comparative similarity, one must be more specific about the nature of mental states. One important example is the theory of Alchourrón, Gärdenfors, and Makinson (AGM) [1, 16], who laid down axioms intended to characterize the observable properties of a notion of logically rational belief revision. Their axioms for contraction are as follows.

($\dot{-}1$) $A \dot{-} x$ is a theory whenever A is
($\dot{-}2$) $A \dot{-} x \subseteq A$
($\dot{-}3$) If $x \notin Cn(A)$, then $A \dot{-} x = A$
($\dot{-}4$) If $\not\vdash x$, then $x \notin Cn(A \dot{-} x)$
($\dot{-}5$) If $\vdash x \leftrightarrow y$, then $A \dot{-} x = A \dot{-} y$
($\dot{-}6$) $A \subseteq Cn((A \dot{-} x) + x)$ whenever A is a theory
($\dot{-}7$) $(A \dot{-} x) \cap (A \dot{-} y) \subseteq A \dot{-} (x \wedge y)$ whenever A is a theory
($\dot{-}8$) If $x \notin A \dot{-} (x \wedge y)$, then $A \dot{-} (x \wedge y) \subseteq A \dot{-} x$ whenever A is a theory

These axioms mainly state fairly intuitive conditions, for example, that $A \dot{-} x$ is always included in A ($\dot{-}2$); that contraction leaves A unchanged if $x \notin Cn(A)$ ($\dot{-}3$); that the contraction omits x as long as $\not\vdash x$ ($\dot{-}4$); that contractions by equivalent statements yield the same result ($\dot{-}5$); and that adding x to the result of contracting by x only yields conclusions present prior to the contraction ($\dot{-}6$).

AGM showed that a range of natural revision functions based on ideas similar to Rescher's preference-based revisions satisfy all their axioms, with different types of AGM revisions corresponding to the way revision proceeds when there are multiple maximal consistent subsets over which Rescher revision preferences operate. In fact, AGM revision implicitly embodies one revision preference also implicit in Rescher's maximize-first approach, namely a uniform preference for believing more over than believing less [10]. This more-is-better preference underlies traditional thought in epistemology, but is better avoided in nonmonotonic reasoning, in which the reasoner bases inference on preferences about whether particular additional beliefs are better than others, or are better than withholding judgment. Both of these theories, then, are reasonably viewed as theories of rational resistance to change.

One can also broaden the AGM conception to apply to revision of beliefs, preferences, and other attitudes [9]. By replacing ordinary logical relations between sentences with the entailment and consistency notions of a Scott information system [33], one can consider AGM-style revisions over any type of mental attitude or entity.

3.3   Collective Versus Component Comparisons

The AGM axioms characterize changes of entire theories, but do not explicitly relate changes affecting different beliefs. Gärdenfors and Makinson [17] studied theory revision from the perspective of individual statements and showed that the outcomes-based AGM approach implies the existence of an ordering of beliefs by relative *epistemic entrenchment*, such that the effect of adopting a new belief is to abandon the least entrenched prior beliefs necessary to ensure consistency. The ordering $x < y$ of two statements in a theory means that one gives up x before y in contractions when one cannot keep both. They formulate the following set of axioms stating that the entrenchment ordering must be transitive (≤ 1); exhibit a general type of reflexivity (≤ 2); must give up at least one conjunct when giving up a conjunction (≤ 3); must regard propositions not in the current beliefs to be minimally entrenched (≤ 4); and must regard propositions to be maximally entrenched only if they are logically valid (≤ 5).

(≤ 1)   If $x \leq y$ and $y \leq z$, then $x \leq z$
(≤ 2)   If $x \vdash y$, then $x \leq y$
(≤ 3)   Either $x \leq x \wedge y$ or $y \leq x \wedge y$
(≤ 4)   If A is consistent, then $x \leq y$ for all y iff $x \notin A$
(≤ 5)   If $x \leq y$ for all x, then $\vdash y$

Gärdenfors and Makinson then prove that the axioms for contraction and entrenchment are equivalent in the sense that every contraction function corresponds to some entrenchment ordering, and every entrenchment ordering corresponds to some contraction function. Formally, $x \leq y$ holds iff either

$$x \notin A \dotdiv (x \wedge y) \quad \text{or} \quad \vdash x \wedge y \; ; \tag{14}$$

correspondingly,

$$y \in A \dotminus x \tag{15}$$

holds iff

$$y \in A \quad \text{and either} \quad x < x \wedge y \quad \text{or} \quad \vdash x \; . \tag{16}$$

Because of their reflexive, transitive, and complete character, entrenchment orderings can also be viewed as preference orders [10]. In this setting, utility or loss functions that represent these preference orders can also be taken as measures of resistance to changing individual beliefs.

4 Substantial Theories of Mental Change

The Lewis, AGM, and Rescher conceptions provide a framework for formalizing comparisons between changes that let us say that one change is bigger than another. Although one expects to find such comparisons in any approach to understanding habit and refraction, mere comparisons do not provide insight into why one change ranks bigger than another, nor do they provide cardinal measures to quantify just how much bigger one change is than another. To understand why one change is harder than another, and just how much harder it is, we must look beyond mere comparisons to theories that explain difficulty in terms of the substance and organization of mental states. To do this, we again turn to Quine, who painted a picture of substantial origins for mental change in his image of the "web of belief" [30, 31], in which belief revision restructures a network of connections between beliefs, with the resulting changes guided by the principle of minimum mutilation.

One key element of the Quinian conception is that connections between beliefs reflect some type of *coherence* between the beliefs. Coherence in this conception demands at least consistency between beliefs, and in some cases that some beliefs explain others, but in all cases that the overall set of beliefs is the central focus of attention. Yet Quine's picture also carries within it another conception, one in which it is direct connections between beliefs that take center stage. Think of a physical web, whether that of a spider or a fisherman's netting, and one thinks of pokes and tugs moving and stretching smaller or larger portions of the web, and of the elements of the web exhibiting degrees of stiffness, elasticity, and the like. Such an approach to mental change has been explored in artificial intelligence in two ways: *structural* or *foundational* methods based on reasons and dependencies and the foundations they provide for beliefs and other attitudes, and *preferential* methods based on domain-specific preferences among beliefs and how they guide rational reasoning.

4.1   Structural Theories of Mental Change

In foundational organizations for mind, the state of belief exhibits a division of memory into at least two parts, variously identified as axioms and theorems [23], reasons and conclusions [3], implicit belief and explicit belief [21], or constructive belief and manifest belief [8]. The essential commonality in all these conceptions is that some type of base elements generate or provide the foundation for some set of extension elements in the sense that beliefs appear as conclusions only if supported by the base or foundational beliefs.

Truth maintenance systems or, more accurately, reason maintenance systems (RMS) and related dependency-based revision frameworks [3, 35, 36] form the exemplars of foundational psychologies. In such systems, records called *reasons* or *justifications* represent finite traces of past derivations or reasoning steps. Traces of ordinary deductive inferences can be represented by deductive reasons

$$A \Vdash C \, , \tag{17}$$

read "A gives C", and standard nonmonotonic inferences can be represented by nonmonotonic reasons

$$A \setminus\!\setminus B \Vdash C \, , \tag{18}$$

read "A without B gives C", informally interpreted as indicating that holding every member of the set A of *antecedent* statements without holding any of the *defeater* statements in B gives each of the *consequence* statements in C. For example, one can represent an inference of the form

> "Conclude (c) Sasha can fly
>   whenever it is believed that (a) Sasha is a bird, and
>   it is not believed that (b) Sasha cannot fly."

by the reason

$$\{a\} \setminus\!\setminus \{b\} \Vdash \{c\} \, , \tag{19}$$

and an "axiom" that Sasha is a bird can be represented by the reason

$$\varnothing \setminus\!\setminus \varnothing \Vdash \{a\} \, . \tag{20}$$

The typical RMS uses these inferential records to construct a set of conclusions that contain all and only the conclusions demanded by the recorded reasons, so that every belief in the set of conclusions is a consequence of some reason for which the antecedents are present (*In*) and the defeaters are absent (*Out*). Formally, a reason

$$A \setminus\!\setminus B \Vdash C \tag{21}$$

requires that the set X of extended conclusions satisfies the condition

$$[A \subseteq X \subseteq \overline{B}] \rightarrow [C \subseteq X] , \tag{22}$$

where $\overline{B}$ denotes the set of all statements not in B. We say that

$$A \setminus\!\setminus B \Vdash C \tag{23}$$

is *valid* if

$$A \subseteq X \subseteq \overline{B}. \tag{24}$$

The requirement that each conclusion be supported by some valid reason, which one might call local grounding, represents one position on a spectrum of possible grounding conditions [5, 12]. The original RMS imposed a strong global grounding requirement, namely that each conclusion is supported by a noncircular argument from the base reasons [3]. One can also consider intermediate organizations that divide memory into multiple working sets or locales and require strict grounding of conclusions within each locale but only local grounding across locales [13].

In a foundational approach, one effects changes in conclusions indirectly by making changes in the base beliefs. One adds reasons that generate beliefs rather than simply adding beliefs directly. This requires systematic addition of reasons in the course of reasoning, for example, through comprehensive recording of the reasoning history [19] or using episodic summarization methods such as chunking [20]. The reasons persist in memory until explicitly removed, even as the set of conclusions changes due to changes in the set of base reasons. In fact, by employing reason structures that provide for uniform nonmonotonic defeasibility, one need not ever remove reasons, but instead can defeat any reason marked for removal [4].

In the foundational setting, reasoned contraction involves tracing underlying derivations recursively. To remove some conclusion, one looks to its supporting reason or argument and seeks to defeat some assumption or reason involved in this support, as in the technique of dependency-directed backtracking [3, 35]. The removal of conclusions by support tracing need not be immediate, because when one assumption is removed or defeated, one must reevaluate the reasons by which it supported other statements. If those supported statements have alternative means of support in other reasons, one must repeat the support-tracing process and make further changes until one finally removes the target conclusion.

4.2 Measuring Revision Difficulty

Reason-based revision, as with any concrete computational method, permits analysis of the difficulty of effecting changes in computational terms of numerical measures of the quantities of time and memory required. Such measures have limited appeal as methods of judging difficulty of change, for the wide variability of time and space requirements due to variations in algorithms and architectures means that the difficulty of a particular change has little relation to specific numerical measures of time and space.

Rather than looking to standard computational measures of difficulty, one can instead employ noncomputational comparisons of difficulty of change related more directly to the reason maintenance approach itself [5]. Such comparisons can take the form of Lewis-like relations

$$X \preceq_W Y \tag{25}$$

of comparative similarity of belief states W, X, Y that capture the intent, if not the practice, of reason maintenance revisions. For example, one can compare change difficulty in terms of the set of changed conclusions, so that

$$X \preceq_W Y \quad \text{iff} \quad X \triangle W \subseteq Y \triangle W , \tag{26}$$

where $X \triangle Y$ represents the symmetric difference

$$(X \setminus Y) \cup (Y \setminus X) . \tag{27}$$

One can also measure change difficulty in terms of the number of changes made, in which case

$$X \preceq_W Y \quad \text{iff} \quad |X \triangle Y| \leq |Y \triangle W| . \tag{28}$$

4.3   Exploiting Substantive Measures in Phenomenal Theories

Although phenomenal theories of mental change provided only qualitative comparisons of difficulty of change, one can connect the substantial and phenomenal theories and seek to transfer cardinal measures from substantial to phenomenal theories.

One can interpret psychological systems as reflecting reasons without any implication that the system actually uses representations like reasons in its operation [5, 12]. In this approach, one examines the range of states exhibited by the system to identify associations between elements of these states. Reasons correspond to fairly simple conditional and nonmonotonic associations between mental elements. With such interpretations, one can use sets of reasons as a fairly general way of describing the origins of degrees of difficulty of change [7].

More direct connections come from reading entrenchment orderings out of foundational structures and vice versa. Entrenchment orderings arise naturally in foundational psychologies. If one draws a graph in which edges labeled by reasons connect antecedent beliefs with their reasoned conclusions, one can assign levels to the assumptions introduced by nonmonotonic reasons. Assumptions with no antecedents occur at the base, and assumptions with antecedents of a certain level occur at the next level. Following these inferential connections produces an assumption graph that shows which assumptions depend on others. The entrenchment rank of a belief then corresponds roughly to its depth in the assumption graph. Base beliefs are the most entrenched, and to a first approximation beliefs closer to the base beliefs or with multiple derivations are more entrenched than beliefs further from the base or with only one derivation.

One can also go the other way, as Gärdenfors [18] suggests, and read reasons out of entrenchment order. Indeed, del Val [39] proved the equivalence of the coherence and foundations approaches under the assumptions that reasons are beliefs (no nonmonotonic rules), that reasons generate conclusions by deductive closure, that logically equivalent conclusions have identical bases, and that one has an entrenchment ordering on foundational beliefs. This formal equivalence has limited practical import, however, because realistic psychologies do not satisfy these idealizing assumptions. Computational embodiments, for example, rely on finite representations and computations, and neurophysiological embodiments involve large but finite numbers of neurons with relatively small fan-in/fan-out connectivity. Beyond these finiteness considerations, there are practical reasons to regard entrenchment as derived from reasons rather than the reverse [11].

4.4 Preferential Theories of Mental Change

One can connect reasoned foundations with entrenchment in another way as well. As noted earlier, one can interpret nonmonotonic reasons as constraints on sets of beliefs, mandating that certain conclusions be held if certain antecedents are held and others are not. In this interpretation, one can regard reasons as expressing intentions of the reasoner about the content of its mental state, namely that conclusions of valid reasons must be believed.

However, one can also interpret nonmonotonic reasons as expressing preferences about the content of its mental state as well [5, 6], somewhat akin to Rescher's revision preferences and the preference-order character of the epistemic entrenchment relation. The preferential content of reasons indicates that indicated nonmonotonic conclusions should be believed in preference to indicated nonmonotonic qualifications, in the sense that sets of conclusions X satisfying

$$[A \nsubseteq X] \vee [A \subseteq X \subseteq \overline{B} \wedge C \subseteq X] \tag{29}$$

are preferred to conclusions satisfying

$$A \subseteq X \nsubseteq \overline{B} , \tag{30}$$

which in turn are preferred to conclusions satisfying

$$[A \subseteq X \wedge X \not\subseteq \overline{B} \wedge C \not\subseteq X] . \tag{31}$$

The interpretation of reasons as expressing preferences about mental states implies that grounded sets of nonmonotonic conclusions are Pareto optimal, that is, satisfy as large a set of reasoning preferences as possible [6]. It also ties conflicting intuitions about the range of possible revision methods to the problems of group decision making [15].

4.5   Substance and Origins of Change

The substantive theories of mental change go beyond the ideals of pure logical and economic rationality by relating the shape of mental changes to the origins of mental changes. The AGM theory of belief revision, for example, says nothing about how the entrenchment order changes, only that it exists and guides revisions, although some subsequent theories treat some revisions as shifting the base of the entrenchment ranking. Similarly, standard theories of economic rationality say much about how probabilistic expectations change with new evidence, but almost nothing about how or when preferences change. Artificial intelligence addresses these questions with theories of problem solving in which reducing goals to subgoals changes the preferences of the reasoner, in which chunking or related mechanisms change the set of base reasons and hence the foundations of mental states, and in which base and derived reasons express preferences that shape the revisions they induce.

5   A Mechanical Perspective

Even though one can interpret a variety of systems in terms of reasons, reasons still represent only one means for assessing difficulty of change. To find a more general identification of generators of difficulty in change, we look to mechanics and its notions of mass, inertia, and constitutive resistive forces. In the present discussion, the base reasons of the foundations approach constitute the mass of the reasoner, the derived conclusions constitute its position, changes in conclusions constitute its velocity, and changes in base reasons constitute its change in mass. Reasoning in such reasoners generates forces that depend on the character of the reasoner, on how the reasoner conducts volitional and deliberative processes.

Everyday references to habit, refraction, and entrenchment make informal use of mechanical notions in addition to the neutral and psychological language used above. In this long-standing but informal usage, refraction becomes reaction to force, and entrenched habits require force to change. This usage probably predates the development of mechanics as a formal, mathematical subject, but ever since the Newtonian era, application of mechanical concepts to anything but physical systems has involved only metaphor and analogy.

Recently, however, axiomatic work in rational mechanics and formal work in artificial intelligence have laid the groundwork for a change in the theoretical status of psychological application of mechanical concepts. Artificial psychological systems such as RMS satisfy the axioms of mechanics, and thus constitute systems to which mechanical concepts properly apply without recourse to metaphor or analogy [14].

Modern rational mechanics [26, 27, 37] supports applications to psychology by providing a formal theoretical structure that offers several advantages over the older mechanical tradition. Three of these advantages offer special benefit in formalizing psychology in mechanical terms.

First, rational mechanics cleanly separates general mechanical requirements from the properties of special materials. Modern theories of mechanics regard bodies as subject to *general* laws applying to all types of materials, laws that set out the properties of space, time, bodies, and forces, and that relate forces on bodies to the motions caused by the forces.

None of the central mechanical laws say anything about which forces exist, or even that *any* particular forces exist. Such statements instead come in *special* laws of mechanics, such as laws of *dynamogenesis* that characterize the origin of forces. The most general of these set out the laws of inertial forces. Traditional presentations of mechanics give a distinguished role to this force through the Newton–Euler equation

$$f = \dot{p} \,, \tag{32}$$

in which one takes the quantity f to aggregate all the non-inertial forces acting on the body. The more general law

$$f - \dot{p} = 0 \tag{33}$$

of the balance of forces places the inertial force $(-\dot{p})$ generated by the mass of the body on an equal footing with other forces, and has, as its fundamental principle the balance of all forces acting on the body, the meaning that all these forces sum to zero.

Other special laws characterize the behavior of special types of materials, ordinarily identified in terms of constraints on bodies, configurations, motions, and forces. For example, one obtains rigid body mechanics from the general theory by adding kinematical constraints that fix the relative distances of body parts; elastic body mechanics comes from adding the assumption that body deformations generate elastic forces dependent on the deformation; and the mechanical theory of rubber comes from modifying the general theory of elastic materials with the configuration-dependent forces characteristic of rubber. Mechanics uses the term *constitutive assumptions* to refer to special laws for particular materials, since each such law reflects an assumption about the constitution of the material. Mechanical practice depends critically on these special laws. Some so-called fundamental laws of physics constitute special laws of "elementary" particles and fields; most of these, however, stand largely irrelevant to the more numerous special laws characterizing ordinary materials. Rigorous derivation of most special laws from "fundamental" properties of elementary particles remains well beyond present theoretical capabilities, even if such derivations exist.

Second, rational mechanics separates much of the structure of ordinary physics from assumptions about continuity of the quantities involved. The modern axioms of mechanics state fundamental properties of mechanical notions with reference only to some basic algebraic and geometric properties of the spaces involved. Ordinary numbers exhibit these properties, but modern algebra and geometry show how many of the familiar properties of numbers required for mechanics also occur in discrete and finite structures. The resulting algebraic structures for space, mass, and force have much the same character as standard conceptions, although the broadened mechanics allows the possibility of different types of mass, much as pre-relativistic classical mechanics regarded inertial and gravitational mass as different mechanical properties that mysteriously had proportionate values. This broadening of the notion of mass means that mass and velocity sometimes combine in different ways than in traditional mechanics.

One can thus purge the axioms of mechanics of continuity assumptions in the same way one purges them of constitutive properties of special materials, and so obtain a mechanics covering discrete and continuous systems, either separately or in combination, in which the usual continuity assumptions need hold only for the usual physical subsystems.

Third, rational mechanics provides a formal characterization of the notion of force in a way that covers psychological notions. Mechanics did not provide any axiomatic characterization of force until the middle of the past century, well after Hilbert made formalization of all of physics one of his famous problems in 1900. As formalized by Noll [26], the theory of force takes on a generality that covers both physics and psychology. The general laws of forces, for example, state the additivity, balance, and frame-indifference of forces on each body. These state that the force on a body is the sum of all the forces on its disjoint subbodies (additivity); that all forces acting on a body add to zero (balance); and that the true force on a body does not depend on the observer (frame-indifference).

Obtaining the full benefits of mechanical concepts in psychology will require considerable mathematical work to provide an analysis of discrete and hybrid mechanical systems that matches the theoretical and methodological power of modern continuum mechanics, as well as work to embed mechanical concepts in languages for specifying and analyzing designs for reasoners.

6 Effort in Discrete Cognition

To illustrate the mechanical perspective on mind, we sketch a highly simplified illustration based on RMS. In this illustration we regard cognitive states as including sets of discrete beliefs, preferences, desires, and intentions, as well as reasoning rules or habits, especially rules of a sort we will regard as reasons or justifications. We write $\mathcal{D}$ to denote the set of all possible attitudes making up cognitive states, so that $\mathcal{P}(\mathcal{D}) = 2^{\mathcal{D}}$ represents all possible states $S \subseteq \mathcal{D}$.

6.1 Kinematics and Dynamics

We represent cognitive states using vectors in the binary vector space $\mathbb{D} = (\mathbb{Z}_2)^{\mathcal{D}}$ over scalars $\mathbb{Z}_2$. In the RMS terminology, 1 means *In* and 0 means *Out*. We define

$$\mathbf{0} = \varnothing = (0, 0, \ldots), \quad \mathbf{1} = \mathcal{D} = (1, 1, \ldots), \tag{34}$$

and

$$\mathbf{1} - x = \overline{x} = \mathcal{D} \setminus x . \tag{35}$$

Addition corresponds to symmetric difference, so

$$x + x = \mathbf{0} \tag{36}$$

and

$$x - y = x + y . \tag{37}$$

Pointwise multiplication corresponds to intersection, so that

$$xy = x \cap y . \tag{38}$$

For infinite $\mathcal{D}$, we consider only vectors representing the finite and cofinite subsets. In this space, orthogonal transformations are permutations.

We write $x_t \in \mathbb{D}$ to denote the position at discrete instant t and

$$\dot{x}_t = x_t - x_{t-1} \tag{39}$$

to denote the trailing velocity. This trailing velocity corresponds to the change signals used in some automated reasoners in triggering rules [3]. The leading acceleration

$$\ddot{x}_t = \dot{x}_{t+1} - \dot{x}_t \tag{40}$$

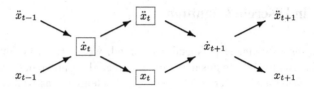

Fig. 1 The kinematical relationships among position variables in time. The *boxed* quantities denote conventional labels for the quantities of interest at instant t, with a reasoning agent observing x_t and $\dot{x}_t$ and choosing $\ddot{x}_t$

reflects the additions and removals indicated by steps of reasoning. We depict these kinematical quantities in Fig. 1.

We denote the mass at an instant by $m_t \in \mathbb{D}$, and the leading mass flux

$$\dot{m}_t = m_{t+1} - m_t \; . \tag{41}$$

The momentum

$$p_t \in \mathbb{D} \times \mathbb{D} \tag{42}$$

decomposes into separate mass and velocity components as $p_t = (m_t, \dot{x}_t)$. We write the leading momentum change as

$$\dot{p}_t = p_{t+1} - p_t = (\dot{m}_t, \ddot{x}_t) \; . \tag{43}$$

We denote the force at an instant by

$$f_t \in \mathbb{D} \times \mathbb{D} \; . \tag{44}$$

Euler's law of linear momentum then takes the familiar form

$$f_t = \dot{p}_t = (\dot{m}_t, \ddot{x}_t) \; . \tag{45}$$

The total force

$$f_t = f_t^a + f_t^s \tag{46}$$

on the reasoner combines the applied force f_t^a of environment on the reasoner with the self-force f_t^s of the reasoner on itself.

In this setting, it is natural to interpret reasons as conditional invariants of the motion. In particular, an *interval reason*

$$r = A_r \, \| \, B_r \, \Vdash \, C_r \, \| \, D_r \; , \tag{47}$$

read "A without B gives C without D", or *antecedents* without *defeaters* gives *consequences* without *denials*, adds its consequences and removes its denials if the reason is *valid* (antecedents *In* and defeaters *Out*).

We obtain the forces generated by such reasons as follows. In the indirect change setting of the RMS, changes to position start with the application of a *mass* force $f_t^a = (\dot{m}_t, 0)$ that adds or removes base reasons. To update the position according to a new base reason r, the RMS then generates a *spatial* self-force

$$f_r^s(x_t, m_t, \dot{x}_t) = (0, \ddot{x}_t) . \tag{48}$$

The application of r produces the new position

$$x_{t+1} = \begin{cases} x_t + C_r \overline{x}_t + D_r x_t & \text{if} \quad A_r \overline{x}_t + B_r x_t = 0 \\ x_t & \text{otherwise} \end{cases} \tag{49}$$

from which we obtain the velocity

$$\dot{x}_{t+1} = \begin{cases} C_r \overline{x}_t + D_r x_t & \text{if} \quad A_r \overline{x}_t + B_r x_t = 0 \\ 0 & \text{otherwise} \end{cases} \tag{50}$$

and acceleration

$$\ddot{x}_t = \begin{cases} C_r \overline{x}_t + D_r x_t - \dot{x}_t & \text{if} \quad A_r \overline{x}_t + B_r x_t = 0 \\ \dot{x}_t & \text{otherwise} \end{cases} \tag{51}$$

6.2 Power and Work

With reasoning mediated by reasons in this way, we can calculate the work and effort expended in reasoning as follows. The *power*

$$P_t = \dot{x}_{t+1} \cdot f_t \tag{52}$$

exerted across interval $(t, t + 1)$ is found in the inner product of the force acting across that interval with the velocity across that same interval. The differing temporal subscripts of velocity and force in this formula reflect the difference between leading forces and trailing velocities. We calculate the instantaneous power P_t to be

$$P_t = \dot{x}_{t+1} \cdot f_t \tag{53}$$

$$= \dot{x}_{t+1} \cdot (\dot{m}_t, \ddot{x}_t)$$

$$= |(\dot{m}_t, \dot{x}_{t+1} \ddot{x}_t)| \tag{54}$$

$$= |(\dot{m}_t, \dot{x}_{t+1} - \dot{x}_{t+1}\dot{x}_t)|$$

$$= |(\dot{m}_t, \dot{x}_{t+1})| - |(0, \dot{x}_{t+1}\dot{x}_t)|, \tag{55}$$

here using the norm that counts the number of 1s in a vector corresponding to the ordinary inner product of binary vectors. The work expended across some interval, therefore, is the integral (sum) of the power over that interval, and we may use this mechanical work as a measure of the mental effort of reasoning.

We examine this formula in greater detail to understand how this measure of effort works in the context of reasoning. Different reasoners operate in different ways, and we find that the overall effort of reasoning varies accordingly.

One can divide reasoning models into two types that differ in the treatment of time. In what one might call the *internal time* model, one identifies instants with steps of reasoning, no matter how long separates these steps in the world at large. In what one might call an *external time* model, one regards steps of reasoning as separated by intervals during which the reasoner does nothing. Both of these models offer useful insights.

We consider the internal time model first. This model corresponds best to a notion of deliberate reasoning, in which every step of reasoning involves some change to memory or outlook. In (55) we see that the power expended across a basic unit of time is the change of mass and position minus a cross term $\dot{x}_{t+1}\dot{x}_t$ involving velocity at successive intervals. This cross-term vanishes in deliberate reasoning because normally one does not immediately retract a conclusion one has just drawn, or draw a conclusion one has just retracted; there would be no point to it. In this setting, therefore, we obtain the magnitude of the power expended by the step of reasoning by

$$P_t = |(\dot{m}_t, \dot{x}_{t+1})| \tag{56}$$

$$= |\dot{m}_t| + |\dot{x}_{t+1}|. \tag{57}$$

In this case, the work of a step of reasoning just adds together the number of changes made in memory and attitudes, so the effort involved in a chain of reasoning steps consists of the cumulative number of changes made in memory and attitudes across the span of reasoning.

In the external time model, we regard steps of reasoning as exerting impulse forces on the reasoner, with the reasoner exhibiting inertial (force-free) motion between steps of reasoning. The "Simple Impulse" table in Fig. 2 illustrates the application of a simple impulse spatial force akin to the internal-time model just discussed. This impulse expends an effort of $|\alpha|$ in the time step in which it is applied, according to the preceding calculation. In the subsequent inertial motion, of course, the force f_t vanishes, and so by (53) the power vanishes as well, so the total effort expended in a chain of reasoning steps again equals the cumulative sum of the number of changes to memory and attitudes, with the inertial motions doing no work.

	Time step	Simple Impulse				Up and Down			
Fig. 2 Kinematic quantities $(x, \dot{x}, \ddot{x})$, power (p), and total effort in two forms of spatial impulse motion starting from rest at the origin location	t	x	$\dot{x}$	$\ddot{x}$	P	x	$\dot{x}$	$\ddot{x}$	P
	0	**0**	**0**	α	$\lvert\alpha\rvert$	**0**	**0**	α	$\lvert\alpha\rvert$
	1	α	α	**0**	0	α	α	α	0
	2	**0**	α	**0**	0	α	**0**	**0**	0
	3	α	α	**0**	0	α	**0**	**0**	0
	4	**0**	α			α	**0**		
	Work	$\lvert\alpha\rvert$				$\lvert\alpha\rvert$			

Inertial motion takes a cyclic form in the discrete space $\mathbb{D}$ due to the algebraic characteristic that $x + x = \mathbf{0}$. As Fig. 2 indicates, inertial motion with velocity α starting from a position α thus traverses the trajectory $\alpha, 0, \alpha, 0, \ldots$. It is certainly not commonplace to think of reasoners as cycling the last set of conclusions in this way. In standard artificial intelligence mechanizations, one instead regards step of reasoning as changing the set of conclusions from one set to another and then leaving it there until the next step of reasoning, as in the internal-time picture of motion. Accommodating this expectation requires one to modify the simplistic picture of reasoning seen in the internal time model.

One easily obtains a more familiar picture of reasoning by regarding steps of reasoning as exerting two impulses, corresponding to the rising and falling edges of a pulse, as depicted in the "Up and Down" table of Fig. 2. That is, the force of the first half of a step of reasoning changes the velocity so as to effect the desired change of position, and the force of the second half of the step of reasoning changes the velocity back to zero by simply reversing (repeating) the thrust of the first half. This produces a pattern of motion of start–stop steps separated by zero-velocity intervals. This start–stop pattern of forces is in fact the pattern of reason forces, in which the frictional force component—$\dot{x}$ provides the falling impulse. This does not involve twice the mechanical effort of the internal time and simple external time pictures, however, because the falling impulses, matched with zero velocities in (54), contribute nothing to the cumulative effort.

Note that mechanical work only measures the effort of making the change itself and does not include any effort involved in evaluating the applicability of some reasoning habit, of searching for the appropriate inference to perform, or of deciding among alternative inferences, if these activities are not effected by means of reasoning steps themselves. If generating a force α requires effort $\lvert\alpha\rvert$, for instance, then the Simple Impulse motion involves an additional effort of $\lvert\alpha\rvert$, while the Up and Down motion involves an additional effort of $2\lvert\alpha\rvert$. The effort associated with such activities, however, depends on the organization and realization of the mind. For example, there need be no effort involved in producing the falling impulse of Up and Down as this value is already available as the velocity. Or for another example, evaluating the applicability of a set of reasons by a serial scan of databases of reasons and conclusions likely involves more effort than by parallel evaluations conducted by reasons wired together like neurons or Boolean circuits.

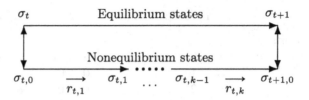

Fig. 3 Reasoned decomposition of a transition between equilibrium states σ_t and σ_{t+1} into a series of "microtransitions" between microstates $\sigma_{t,i}$ effected by a series of reason applications $r_{t,i}$

6.3   Refraction and Elasticity

The effort of reasoning also involves the work of focusing attention on reaching the intended conclusions and on not pursuing distractions arising in the course of reasoning. Autonomous habits of reasoning or perceptions can exert forces of distraction on the reasoner and forces that move it either back toward where it started or merely off the intended path. The reasoner must notice these backwards steps and diversions and force itself back onto the intended path. One pays a double price for refraction and distraction, as both these and the refocusing actions do work and add to the total effort of reasoning. The same holds true for switching between separate reasoning activities being pursued simultaneously, for the effort of switching contexts adds to the efforts attributable to the separate reasoning activities.

The effort involved in a single step of reasoning also enters into the comparisons made by coherence-based conceptions of mental change indirectly because these large-scale changes to mental state typically involve an extended process of smaller reasoning steps. Mechanically, the coherence-based changes take the form of elastic response. The stipulated change, or step of reasoning that triggers a change to a new coherent state of mind, consists of an external deformation. This deformation requires a change to a new "relaxed" or equilibrium state. In RMS revision, this relaxation process consists of a sequence of smaller reasoning steps, which we depict in Fig. 3. As the base reasons constituting the mass of the reasoner grows through learning and experience, the time needed to effect revisions can also grow as revisions involve more and more reasons.

Nonmonotonic reasoning exhibits an elastic character even more strongly. One can view the defeat of an otherwise valid nonmonotonic reason as producing a deformed configuration of the reasoner. Defeat or removal of this defeater forces restoration of the state of mind prior to imposition of the deformation, with the nonmonotonic reason acting like a spring element in shaping the mental configuration. The character of such elastic response is more complex in reasoning than in simple physical materials like springs in that RMS-based reasoners search for equilibria in both relaxation and restoration processes, and the equilibrium resulting from a restoration need not coincide with the one existing prior to the deformation. One must look at more complicated physical systems than individual springs to see similar indeterminacy of equilibrium transitions.

7 Conclusion

Modern rational mechanics provides concepts of mass, force, work, and elasticity that constitute a new analytical framework and vocabulary for characterizing limits to mental change. These concepts go beyond those available in theories of ideal logical and economic rationality that focus on mere comparative difficulty, and beyond computational and neurophysiological theories that focus on measures only loosely related to mental effort. Mechanics also provides other means for characterizing limitations on mental change, including bounds on the forces exerted in reasoning and forms of constitutional rigidity (see [14]).

References

1. Alchourrón, C., Gärdenfors, P., Makinson, D.: On the logic of theory change: partial meet contraction functions and their associated revision functions. J. Symbolic Logic **50**, 510–530 (1985)
2. Brooks, R.A.: A robust layered control system for a mobile robot. IEEE J. Robot. Automat. **2**(1), 14–23 (1986)
3. Doyle, J.: A truth maintenance system. Artif. Intell. **12**(2), 231–272 (1979)
4. Doyle, J.: A model for deliberation, action, and introspection. AI-TR 581, Massachusetts Institute of Technology, Artificial Intelligence Laboratory (1980)
5. Doyle, J.: Some theories of reasoned assumptions: an essay in rational psychology. Technical Report 83–125, Department of Computer Science, Carnegie Mellon University, Pittsburgh (1983)
6. Doyle, J.: Reasoned assumptions and Pareto optimality. In: Proceedings of the 9th International Joint Conference on Artificial Intelligence, pp. 87–90. Morgan Kaufmann, San Francisco (1985)
7. Doyle, J.: Artificial intelligence and rational self-government. Technical Report CS-88-124, Carnegie-Mellon University Computer Science Department (1988)
8. Doyle, J.: Constructive belief and rational representation. Comput. Intell. **5**(1), 1–11 (1989)
9. Doyle, J.: Rational belief revision. Presented at the third international workshop on nonmonotonic reasoning, Stanford Sierra Camp, CA, June 1990
10. Doyle, J.: Rational belief revision (preliminary report). In: Fikes, R.E., Sandewall, E. (eds.) Proceedings of the Second Conference on Principles of Knowledge Representation and Reasoning, pp. 163–174. Morgan Kaufmann, San Mateo (1991)
11. Doyle, J.: Reason maintenance and belief revision: foundations vs. coherence theories. In: Gärdenfors, P. (ed.) Belief Revision, pp. 29–51. Cambridge University Press, Cambridge (1992)
12. Doyle, J.: Reasoned assumptions and rational psychology. Fundamenta Informaticae **20**(1–3), 35–73 (1994)
13. Doyle, J.: Final report on rational distributed reason maintenance for planning and replanning of large-scale activities (1991–1994). Technical Report TR-97-40, ADA328535, Air Force Research Laboratory (1997)
14. Doyle, J.: Extending Mechanics to Minds: The Mechanical Foundations of Psychology and Economics. Cambridge University Press, London (2006)
15. Doyle, J., Wellman, M.P.: Impediments to universal preference-based default theories. Artif. Intell. **49**(1–3), 97–128 (1991)

16. Gärdenfors, P.: Knowledge in Flux: Modeling the Dynamics of Epistemic States. MIT, Cambridge (1988)
17. Gärdenfors, P., Makinson, D.: Revisions of knowledge systems using epistemic entrenchment. In: Vardi, M.Y. (ed.) Proceedings of the Second Conference on Theoretical Aspects of Reasoning About Knowledge, pp. 83–95. Morgan Kaufmann, Los Altos (1988)
18. Gärdenfors, P.: The dynamics of belief systems: foundations vs. coherence theories. Revue Internationale de Philosophie **172**, 24–46 (1990)
19. de Kleer, J., Doyle, J., Steele Jr., G.L., Sussman, G.J.: AMORD: explicit control of reasoning. In: Proceedings of the 1977 Symposium on Artificial Intelligence and Programming Languages, pp. 116–125. ACM, New York (1977)
20. Laird, J.E., Rosenbloom, P.S., Newell, A.: Chunking in soar: the anatomy of a general learning mechanism. Mach. Learn. **1**(1), 11–46 (1986)
21. Levesque, H.J.: A logic of implicit and explicit belief. In: Proceedings of the National Conference on Artificial Intelligence, American Association for Artificial Intelligence, pp. 198–202. AAAI Press, Palo Alto, CA, USA (1984)
22. Lewis, D.: Counterfactuals. Blackwell, Oxford (1973)
23. McCarthy, J.: Programs with common sense. In: Proceedings of the Symposium on Mechanisation of Thought Processes, vol. 1, pp. 77–84. Her Majesty's Stationery Office, London (1958)
24. Minsky, M.: The Society of Mind. Simon and Schuster, New York (1986)
25. Newell, A., Simon, H.A.: Human Problem Solving. Prentice-Hall, Englewood Cliffs (1972)
26. Noll, W.: The foundations of classical mechanics in the light of recent advances in continuum mechanics. In: Henkin, L., Suppes, P., Tarski, A. (eds.) The Axiomatic Method, with Special Reference to Geometry and Physics; Proceedings of an International Symposium held at the University of California, Berkeley, December 26, 1957–January 4, 1958. Studies in Logic and the Foundations of Mathematics, pp. 266–281. North-Holland, Amsterdam (1958). Reprinted in [28]
27. Noll, W.: Lectures on the foundations of continuum mechanics and thermodynamics. Arch. Rational Mech. Anal. **52**, 62–92 (1973). Reprinted in [28]
28. Noll, W.: The Foundations of Mechanics and Thermodynamics: Selected Papers. Springer, Berlin (1974)
29. Quine, W.V.: Philosophy of Logic. Prentice-Hall, Englewood Cliffs (1970)
30. Quine, W.V.O.: Two dogmas of empiricism. In: From a Logical Point of View: Logico-Philosophical Essays, 2nd edn., pp. 20–46. Harper and Row, New York (1953)
31. Quine, W.V., Ullian, J.S.: The Web of Belief, 2nd edn. Random House, New York (1978)
32. Rescher, N.: Hypothetical Reasoning. North Holland, Amsterdam (1964)
33. Scott, D.S.: Domains for denotational semantics. In: Nielsen, M., Schmidt, E.M. (eds.) Automata, Languages, and Programming: Ninth Colloquium. Lecture Notes in Computer Science, vol. 140, pp. 577–613. Springer, Berlin (1982)
34. Shand, A.F.: The Foundations of Character: Being a Study of the Tendencies of the Emotions and Sentiments, 2nd edn. Macmillan and Company, London (1920)
35. Stallman, R.M., Sussman, G.J.: Forward reasoning and dependency-directed backtracking in a system for computer-aided circuit analysis. Artif. Intell. **9**(2), 135–196 (1977)
36. Sussman, G.J., Stallman, R.M.: Heuristic techniques in computer-aided circuit analysis. IEEE Trans. Circ. Syst. **CAS-22**(11) (1975)
37. Truesdell, C.: A First Course in Rational Continuum Mechanics, vol. 1. Academic, New York (1977)
38. Ullman, S.: Visual routines. AI Memo 723, MIT AI Lab, Cambridge (1983)
39. del Val, A.: Non-monotonic reasoning and belief revision: syntactic, semantic, foundational, and coherence approaches. J. Appl. Non-Classic. Logics **7**(2), 213–240 (1997). Special Issue on Inconsistency-Handling

Semantic Technologies: A Computational Paradigm for Making Sense of Qualitative Meaning Structures

Udo Hahn

Abstract The World Wide Web has become the largest container of human-generated knowledge but the current computational infrastructure suffers from inadequate means to cope with the meaning captured in its interlinked documents. Semantic technologies hold the promise to help unlock this information computationally. This novel paradigm for the emerging Semantic Web is characterized by its focus on qualitative, from the perspective of computers un(der)structured, data, with the goal to 'understand' Web documents at the content level. In this contribution, fundamental methodological building blocks of semantic technologies are introduced, namely terms (symbolic representations of very broadly conceived 'things' in the real world), relations between terms (assertions about states of the world), and formal means to reason over relations (i.e. to generate new, inferentially derived assertions about states of the world as licensed by rules). We also discuss briefly the current status and future prospects of semantic technologies in the Web.

1 Structured Versus Unstructured Data: Quantitative Versus Qualitative Computing

A considerable portion of our knowledge of the world is represented by *quantitative* data and, thus, based on number systems with well-defined formal operations on them, suitable metrics, and measurement units. Our daily life requires us, for instance, to deal with room and outdoor temperatures, the air pressure of the tires of our car, exchange rates of currencies, national gross incomes or employment rates, the likelihood of rainfall today, as well as calory counts of our meals. Such quantitative, metrical data reflect deep scientific abstractions from empirical observables

U. Hahn
Department of German Linguistics Friedrich Schiller University Jena, Jena, Germany
e-mail: udo.hahn@uni-jena.de

B.-O. Küppers et al. (eds.), *Evolution of Semantic Systems*,
DOI 10.1007/978-3-642-34997-3_8, © Springer-Verlag Berlin Heidelberg 2013

studied in physics, economics or medicine that employ the rules of special branches of mathematics, calculus and statistics, in particular. Data generated on this basis have a clear meaning since they rely on well-defined algebraic and metrical criteria and conditions of measurement. Computers as hardware devices and the vast majority of software running on them are typically seen to support numerical computation schemes only involving quantitative data (e.g. solving differential equations, performing matrix multiplications or eigenvalue computations) and the processing of large amounts of numerical data—the reason why computers for decades were also termed 'number crunchers'. A typical computational vehicle to organize and manage large volumes of such data in a thematically coherent way are database systems. They impose additional structure on these data as defined by the underlying organizational structure (e.g. relational schemata) of the database. In this way, quantitative data, in addition, become *structured* data.

Yet, the overwhelming part of our knowledge about the world is not quantitatively represented. It is rather encoded by non-numeric, non-metrical *qualitative* symbol systems whose intrinsic properties, one might suspect, seem to resist mathematical rigor, most prominently everything we express in some form of natural language. The reasons for using natural languages in our everyday and scientific communication are manifold. They are humans' most natural communication vehicle (we all speak at least one language, our mother tongue), and they are so rich in expressivity and flexible to use. Natural languages allow us not only to formulate crisp claims with precise statements but also to add information that overlays these assertions. We may, for instance, establish different temporal axes for an utterance (whether it holds in the past, the presence or the future), or attach different degrees of certainty (with lexical markers such as 'possibly', 'perhaps', etc.) or modality (using 'should', 'may', 'could', etc.) to an utterance. These linguistic means are necessary to formulate intentionally vague assumptions, claims, guesses, opinions or stipulate hypotheses. Using natural language, we may inform, comment on, challenge, criticize, convince or persuade others. This versatile communication power, however, comes with a high potential for ambiguity (i.e. multiple interpretations of the same underlying utterance). Natural languages are adaptive and dynamic systems that undergo continuous changes of the lexical (vocabulary) system and (to a lesser degree though) the grammar system, i.e. the rules guiding the construction (and analysis) of utterances at the syntactic (structural) and semantic (meaning) level. Moreover, as far as linguists can tell, the lexical and grammar systems are very large and interact heavily at all levels of usage and description.

Altogether, these properties of natural languages make it hard for computers to deal with the contents contained in qualitative raw data such as letters, notes, manuals, scientific or newspaper articles, etc. Today's computers primarily allow straightforward text editing and formatting, deleting and inserting character strings at the linear level of so-called *string processing* in text processors. But standard computer programs dealing with textual data as character sequences only have no clue at all what the *contents* of the underlying text really is, e.g. what the facts encoded in it are, whether these facts are true or consistent relative to what is commonly shared knowledge, whether the statements being made are well-argued assertions, unfounded claims, certain or uncertain hypotheses, or merely vague guesses. In a word, the *semantics* of natural language statements concerned with the meaning of single linguistic utterances and their contextual embedding both relative to the discourse setting in which they are uttered and the background knowledge to which they refer currently remain outside the scope of standard computational facilities.

This lack of semantic competence of today's computational machinery is not at all restricted to the understandability of natural language utterances but carries over to other kinds of qualitative symbol systems in which hard-to-interpret sequences of textually encoded symbols, on the one hand, or visual data (such as tables, figures, drawings, pictures, and movies) as well as audio data (such as sounds, music, and spoken language), on the other hand, may constitute meaningful contents. Consider as yet another example the case of the four chemical bases Guanin, Thymin, Cytocin, and Acetin from which DNA sequences are formed. Since many years, powerful string readers (implemented in high-throughput sequencing robots) are available, which read out tons of data constituted by sequences of this 4-letter vocabulary and compare pairs of DNA strings for similarity by checking insertions, deletions, or permutations of characters or character substrings, most notably using variants of the famous BLAST string processing algorithm [1].[1] But only few algorithmic means are available, on a larger systematic scale, to automatically link a syntactically decoded DNA substring to its biomedical meaning or function, e.g. where a piece of a DNA string encodes protein (mal)functions or, even more so, diseases. Obviously, not only the semantics of natural languages but also that of biological ones remain, by and large, *terra incognita* for computing machines.

From the perspective of computers at least, qualitative symbol systems such as natural or bio-molecular languages (for further analogies, see Searls' seminal article [40]) appear as being *unstructured* because computers have no built-in interpretation schemes to decode the intended meaning from sequences of their respective base vocabularies, words and chemical bases, respectively, and the structural configurations they are embedded in according to the underlying syntax. Information contained in such qualitative raw data is, however, at the core of the emerging Semantic Web (Web 3.0)[2] [5].[3] Technologies are being developed to cope with the meaning of this type of data and, thus, determine the structure inherent to qualitative data computationally. The challenge for this undertaking is huge because humans, unlike computers, are particularly proficient to make sense of

---

[1] http://blast.ncbi.nlm.nih.gov/Blast.cgi

[2] http://www.w3.org/2001/sw/

[3] Seth Grimes discusses the myth of the 80% rule (80% of the data on the Web are stipulated to be un- or semi-structured, whereas only the remaining 20% are structured and thus easily interpretable by machines) that has evolved from several hard to track sources (mostly observations made by IT companies, e.g. database, network and storage suppliers); see http://clarabridge.com/default.aspx?tabid=137&ModuleID=635&ArticleID=551. A summary of the arguments related to the proliferation of unstructured data is contained in [41].

such (seemingly) unstructured data, interpret them in context and infer relations between single data items based on previous individual and community knowledge. Interestingly, the semantic interpretation gap we witness not only relates to socially defined and human-rooted symbol systems like natural languages but applies to ones determined solely by nature, like bio-molecular languages, as well.

Why do computers do such a good job when it comes to temperatures, air pressure, trading indicators and physical parameters of nutrition consumption but fail miserably when the semantics of qualitative symbolic systems such as natural or bio-molecular languages come into play? After all, it seems that the meaning of a natural language statement such as 'Jack played tennis with John' is almost impossible to represent by some standard form of calculus statement (as is commonly used for quantitative data), since *prima facie* no formal operation involving summation, multiplication, square roots, integrals or differentiation, etc. plays a role here. Although some aspects of a tennis match can be captured directly by quantitative statements targeting, e.g. the velocity of a tennis ball, the length of the tennis match in time, the oxygen consumption of the players, all these aspects are merely peripheral for an adequate meaning representation capturing the very essence of a tennis playing event. Accordingly, focusing on the relevant actors (the players), the rules of the game, its location and the framing conditions required to play a tennis match outdoors (*versus* indoors tennis) are as important as features like the capabilities and actual performance of the players, decisions of the referee, decisive breaks, weak services and deadly volleys, and the final outcome (who won?). What is seriously lacking to capture these crucial information pieces explicitly is an appropriate formal representation system for such a setting so that computers can 'understand' what's going on (a tennis match, not a football game, not a theater play). What is needed here is a system comparable in mathematical rigor to ones manipulating numerical data, one that can reasonably deal with 'qualitative' rather than (only) quantitative data in a computationally feasible way. Yet, is computation without numbers only wishful thinking, or is it a serious alternative?

Fortunately, there have been efforts already to capture this qualitative slice of reality with formal, computational means. One branch of these activities is rooted in computer science where researchers in the field of artificial intelligence (AI) deal with the representation (i.e. formalization) of common-sense knowledge and corresponding reasoning (i.e. computation) procedures since many decades (see, e.g. [10, 35] or [38]). The fundamental goal of this work is to find a formal foundation to describe the behaviour of human (and virtual) agents acting intelligently, perhaps collaboratively, in the real (and virtual) world, i.e. search, reason, plan, learn and solve problems, but also visually and acoustically perceive information in order to individually experience, reflect on and subsequently change states of the real (or virtual) world. By doing so, such agents converse with other agents (i.e. talk) about the world's past, current and future states. This goal is much broader than the ambitions of linguistic semanticists who, for more than a century, have been studying the relevant entities, their intrinsic properties and the interdependencies among these entities relevant for formal meaning analysis of natural language utterances (see, e.g. [11]). One of the main insights they gained (earlier than AI researchers, by the way) was that a reasonable candidate for *formal* semantic analysis of natural language (though certainly not the only one) were systems based on *formal logics* (for an overview, see, e.g. [6])—several variants of predicate logics (first as well as higher-order) and even more sophisticated formal systems such as modal logics of different types (including temporal, spatial or deontic logics, just to mention a few of the more powerful and thus more expressive systems). Since logical computation, irrespective of the fact whether it is done deductively, inductively, abductively, etc., is completely different from numerical computation, we claim that computers are mostly ignorant of qualitative structure computing because they were (and still are) not directly equipped with computational means to deal with a suitable form of mathematics, formal logics, in particular.[4]

One caveat may be appropriate here. We do certainly not claim that formal semantic computations for qualitative symbolic systems, such as natural languages, can only be carried out within a logical representation framework. Recent efforts in probabilistic semantics (including Bayesian, game and information theoretical approaches) clearly demonstrate that statistical approaches (e.g. co-occurrence statistics, tf-idf metrics, latent semantic analysis), together with the sheer mass of data on the Web [22], also provide a reasonable framework for meaning analysis, in particular for all kinds of classificatory problems (see, e.g. [33]). Hence, we obviously have to dig a little deeper to uncover the essential objects of interest for semantic analysis of meaning structures of natural languages and other qualitative symbol systems.

2 Building Blocks of Qualitative Semantic Computing

In this section, we shall introduce fundamental building blocks for semantic computation. They comprise the very essence almost every formal semantic system for qualitative, symbol-based data is made of, irrespective of whether it is a formal logical or a formal statistical system for semantic analysis. In particular, we will discuss the notions of terms in Sect. 2.1, relations between terms (assertions) in Sect. 2.2. and inference rules in Sect. 2.3, the first two constituting the static (declarative) part, the last one the dynamic (procedural) part of semantic computing (for the latter, no direct correlate exists in statistically based semantic computation schemes though).

---

[4]One of the most remarkable exceptions to this observation is due to the introduction of the logic programming paradigm around 1980 and PROLOG, its primary programming language representative, which was based on Horn clause logic [13].

2.1   Terms

The basic, non-decomposable units of semantic analysis are here called '*terms*'; they refer to the objects (or entities) natural language utterances (but also visual or auditory data streams) are about, without taking into account the way they are combined in terms of assertions (see Sect. 2.2). Considering our example from above ('Jack played tennis with John'), the basic terms contained in this sentence are the players, 'Jack' and 'John', as well as the event they enjoy, namely 'playing tennis.' Still, the notion of a 'term' remains open for further clarification. One prominent distinction for semantically different forms of terms sets concrete individual entities (e.g. 'my friend John' or 'my tennis match this morning') apart from class- or type-level entities (e.g. all persons ever born, or all tennis matches that have been or will ever be played). We may, tentatively, define (the semantics of) the *class* of tennis matches by the union of all *instances* of (all individual) tennis matches. Moving from directly observable instances in the concrete physical world to theoretically constructed types is a major conceptual abstraction from perceived reality that will allow us to further organize the set of all types by means of semantic relations, or, alternatively, assign inherent properties, attributes, conditions, and constraints to them on a definitional basis (see below).

However, defining a class as a set of individual instances might not be fully satisfactory from a semantically 'constructive' point of view since this is an entirely formal approach to term definition. In particular, it does not clarify the intrinsic conceptual properties that allow us to distinguish my tennis match this morning from a particular rose in your garden and our neighbour's car (besides the fact that they constitute different instances and are assigned different, completely unrelated types). What is missing here is a *constructive* semantic specification of a term, either implicitly by stating terminological *semantic relations* it has to other terms (at the class/type level), or explicitly by a *definition* of the meaning of a tennis match, a rose and a car. Such a definition should be exhaustive enough so that any other meaning of a term (at the class/type level) can be excluded from the one being defined.

There are two foundational types of terminological *semantic relations* by which terms can be linked, a taxonomic and a partonomic one.[5] The first one relates a semantically specific to a semantically more general term. In linguistic semantics (see, e.g. [15]), this specific-to-general relation is called *hyponymy* (or general-to-specific *hypernymy*), in the AI knowledge representation community it is called '*Is-a*', and in documentation languages within information science it is called '*narrower*' (more specific) *versus* '*broader*' (more general) term. For example, a 'tennis match' (hyponym) is a specific 'sports event' (hypernym), a 'rose' is a specific 'flower', and a 'car' is a specific 'vehicle'. The generality of the hypernyms can easily be shown since there are sports events other than tennis matches,

---

[5]We distinguish *terminological* semantic relations from *assertional* semantic relations, the latter covering the multitude of empirical assertions such as *interacts-with, is-effective-for, is-mother-of, and is-located-in.*

flowers other than roses and vehicles other than cars. The second terminological relation binds terms indicating a whole to terms indicating the whole's constituent parts. In linguistic semantics, the part-to-whole relation is called *partonymy* (or whole-to-part *holonymy*), the AI knowledge representation and information science communities call it *'Part-of'* (or *'Has-part'*). For example, a (prototypical) 'rose' has as (non-exhaustive) constituent parts 'root', 'leaves', and 'thorns', while a 'car' has 'engine', 'brakes', 'tires' and 'seats' as its parts. Linguistic lexicons make heavy use of such semantic relationships (see, e.g. WORDNET),[6] but also thesauri and terminologies for various science domains are based on these relations (see., e.g. the *Unified Medical Language System* (UMLS) Semantic Network),[7] as well as a bunch of common-sense and domain-specific ontologies (for the life sciences, see, e.g. the *Foundational Model of Anatomy* (FMA)[8] or the *Gene Ontology* (GO)[9]). Although one might think that both relation types can be attributed a fairly straightforward, non-controversial semantics, this is true neither for the taxonomic (see, e.g. [9]), nor for the partonomic case (see, e.g. [39]).

If terms are structured in this way, a semantic network (formally, a directed graph) can be built from such individual terminological relations, where terms form the vertices and terminological semantic relations form the edges of a concept graph. One may navigate in such a graph by following the edges (links) according to certain graph traversal strategies (e.g. moving from a term only following its *'Is-a'* links should lead to more and more general terms; for example, 'rose' *Is-a* 'flower' *Is-a* 'plant' *Is-a* 'eukaryote'). Note that relational encodings of this type constitute an implicit semantic characterization of the meaning of terms because terms are only defined indirectly, relative to other terms in the network. Still, this representation format is already manageable by a computer, since we have mapped a system of unstructured terms to a formal construct (a directed graph) that is directly *representable* in computers (e.g. as a term–term adjacency matrix) and *processable* (e.g. finding fully connected relation paths in that matrix via connectivity search). Perhaps, the most prominent and advanced example of this representation approach is *Conceptual Graphs* [45].

The standard approach to an explicit, constructive textual definition is the Aristotelian type. We specify the *definiendum* (what has to be defined) by a series of properties (*idealiter* an exhaustive list of defining attributes and accompanying constraints), the *definiens*, such that no other term (at the type level) also fulfills the conjunction of all these properties. For instance, we may define 'tennis' as a ball game played by either two or four players involving a yellow ball (of a certain diameter and texture made from a special material). Furthermore, the game is played either on a sand, a lawn or a rubber court (a size-wise well-defined quadrangle separated in the half by a let of a certain height) where each player uses a tennis

---

[6]http://wordnet.princeton.edu/

[7]http://semanticnetwork.nlm.nih.gov/

[8]http://sig.biostr.washington.edu/projects/fm/

[9]http://www.geneontology.org/

racket for playing the tennis ball. Additional rules of the game (e.g. who serves when, from which position on the court, legal ball moves (in- and out-conditions), the scoring of points, winning conditions for a game, a set, and a match, etc.) might be made explicit as well, if necessary. Obviously, this task is truly open-ended since it is hard to decide, if not impossible at all, when the list of properties is really complete.

For human speakers of the same language, with a reasonable level of semantic competence, this indeed helps clarify the meaning of a term. This is so because we have a *genus proximum* ('ball game', a general class of games which subsumes tennis but also other ball games such as football, hockey, basketball), with various *differentia specifica* (e.g. 2 or 4 players, sand, lawn or rubber court, i.e. properties which in their entirety distinguish tennis from all other alternative ball games). However, such a natural language definition cannot be directly used for formal semantic analysis since natural language is not per se a formal language (we will come back to this problem in Sect. 2.2).

Natural language is a poor candidate to serve as a formally adequate semantic meta language for several reasons. Among the most pressing problems encountered are a systematic semantic overloading and excessive proliferation of meaning variants for single terms (and term variants for a single form of meaning, as well). First, very often natural languages provide several alternatives for denoting the same term by different linguistic expressions (e.g. 'car' and 'automobile' are *synonymous* at the monolingual level, while 'car', 'Auto', 'voiture', etc. are synonymous at the multilingual level of several natural languages—they all denote the same semantic type). Second, one term label (its natural language name) may denote different types of entities (e.g. the English term name 'bank' denotes at the same time a river bank, a piece of furniture and a financial institute, and is thus *semantically ambiguous*). Furthermore, definitions are, in principle, required for each natural language term. On the one hand, their number is huge (size counts for comprehensive natural language lexicons enumerate up to 150,000 lexical items for covering the general language portion of a specific natural language,[10] not counting the possible extensions of this lexical set by one or more, e.g. scientific, sublanguages). Even worse, a substantial part of such a lexicon is intrinsically hard to define uncontroversially (consider, e.g. loaded terms such as 'freedom', 'dignity', or 'respect'; or colors such as 'green' 'red' or 'blue'; or moods such as 'happy', 'angry' 'sad' or 'stubborn').

We here face the problem of all-embracing and at the same time often rather complex definition tasks. Sciences in general, quantitatively based ones such as physics or chemistry, in particular, profit from a reductionist view on the world where only a constrained slice of the reality has to be taken care of for theoretical considerations, usually abstracting away from many idiosyncrasies one encounters in reality. For instance, under the particular view of physics or chemistry different

---

[10]This number is based on the lexical population statistics of WORDNET, the most comprehensive lexical database for the English language.

degrees and shapes of rusty materials such as iron are simply of no concern (but theoretical abstractions such as iron's atomic weight and binding constraints relative to other chemical elements are), whereas building and construction engineers have to struggle heavily with such phenomena in their daily professional work. Other kinds of sciences, e.g. jurisprudence, medicine, etc., suffer as well from not being able to reduce reality by definition. Natural language semanticists also face the dilemma that virtually all lexical items of a natural language have to be defined, in principle. Aside from the sheer size and multitude of semantic descriptions, complexity is yet another hard problem for linguistic semanticists. While physics may focus on fundamental notions such as 'mass', 'energy', and 'electrons', from which complex theories are built in a compositional and constructive way, linguists have neither quantitative parameters, nor exhaustive and uncontroversial criteria for defining the semantics of qualitative notions such as 'sunshine', 'smell', 'politics', 'fun' or 'justice'.

Terms have their merits for text analytics in the sense that they characterize what a document is about, i.e. its theme(s) or topic(s). Such a level of analysis, however, does not tell us anything about assertions, i.e. the statements, claims or hypotheses being made in a document, which relate basic entities mentioned in this context so that, e.g. a truth status can be assigned to them (e.g. the assertion 'Jack played tennis with John' is either true or false under well-specified spatio-temporal conditions, whereas 'Jack' cannot be assigned such a truth value at all).

2.2 Relations Between Terms: Assertions

Knowledge is built up in documents when terms are relationally combined to form assertions (predicates) about states of the world. Assertions, at the level of a meta language, have a well-defined structure. An *assertion* is formed from a (usually, nonzero) list of arguments which are typically ordered canonically (argument positions are taken by terms, but more complex variants, functions or even other predicates, are also possible). Each argument may be assigned a semantic class restriction (a so-called argument type) which semantically constrains the term to be taken at that position to this particular type. Finally, all this has to be tied up relative to a specific predicate symbol to which the various arguments are assigned; this symbol characterizes the type of relationship which the arguments share.[11] This format, a predicate symbol plus a list of *n* terms (including functions or predicates), constitutes an assertion, which can formally also be considered as a mathematical *relation*.

---

[11]Formally speaking, we here outline a system which is usually referred to as typed first-order logic (FOL; see, e.g. [18]). Throughout the discussion, we avoid going into low-level details of FOL, in particular, in what concerns the use of quantifiers.

For illustration purposes, consider our tennis match example from above. The sentence 'Jack played tennis with John' describes a particular sports event, an instance of the semantic type *Playing_tennis*. The semantic description of a tennis match requires, among other arguments, players as constituent entities. Players are human actors, such as specific persons, e.g. $John_i$ and $Jack_j$ (the indices i and j denote special instances of all persons whose given name is either *John* or *Jack*, respectively; for the sake of readability, we will leave away such indices whenever possible). We may roughly characterize this assertion as *Playing_tennis* ({*Jack, John*}$_{ACT}$); the (nonstandard) set notation merely indicates the players involved, more generally the ACTors of the tennis match. Of course, this representation is incomplete in many ways; e.g. one may consider the LOCation, the DURation, or even the RESult of the match as being crucial knowledge for a tennis match that has to be specified as well. If we integrate these considerations and insert type-specific placeholders (actually, typed variables) into predicates, we end up with a representation structure like the following one: *Playing_tennis* ({*Jack, John*}$_{ACT}$.___LOC.___DUR.___RES). For an n-ary predicate, we have n argument positions associated with it (so the above predicate is 4-ary, the first argument is already instantiated by two actors, whereas the remaining three argument positions remain uninstantiated so far, i.e. only their type requirements are made explicit). One particular problem we here encounter for semantic descriptions is that, for virtually each predicate one can think of, the number n cannot be fixed *a priori*. This problem resembles one well-known in linguistics, namely the distinction between (syntactically required) complements (e.g. the number and types of grammatically necessary (in)direct objects, the subject, etc.) and (syntactically/semantically possible) adjuncts (e.g. different types of optional spatial, temporal, and manner-type extensions; as an example, consider the extended example sentence 'Jack played tennis with John [on a sand court]$_{LOC-1}$ [in Florida]$_{LOC-2}$ [on a Sunday morning]$_{TIME}$ [in an A-level tournament]$_{MANNER-1}$ [under extremely humid weather conditions]$_{MANNER-2}$ . . .').

The additional information provided by assertions is the composition of terms under a particular predicate name. Several assertions can be combined as sets of assertions. Such a store of knowledge is usually called a *knowledge base*. We may, of course, add or delete assertions from such a knowledge store in the same way as we enter or delete entries in a database. But there are no internal mechanisms to compute *new* assertions (e.g. we may conclude from Jack's and John's playing outdoor tennis that, at the time of playing, no heavy rainfall occurred in the area of the tennis court). Given these considerations, knowledge bases as sets of relation tuples bear a strong similarity to the type of 'tabular' knowledge contained in databases. It might also be worthwhile to emphasize the fact that the transformation of the informal natural language utterance 'Jack played tennis with John' into the formal assertion *Playing_tennis* ({*Jack, John*}$_{ACT}$) constitutes a perfect example for moving from unstructured data to structured ones. Note also that this assertional format, unlike the original sentence, is readily manageable and processable by machine. The distinctive feature, however, that sets knowledge bases apart from databases—their theory-guided extensibility by the application of inference rules— will be discussed in Sect. 2.3.

Besides the open-ended arity of predicates some varieties of knowledge are inherently hard to express in this framework, among them, e.g. uncertain, imprecise, vague or fuzzy knowledge. For instance, sentences such as 'Jack *might* have played tennis with John', '*Usually*, Jack plays tennis with John', '*I guess* Jack played tennis with John', or 'If last Sunday was dry, Jack *should have* played tennis with John' illustrate that our discourse provides lots of superimposed qualifiers for statements which express (dis)beliefs, (un)certainties, etc. related to the base statement, which go far beyond the simple scheme we have been discussing up until now. Indeed, they open up the area of probabilistic logics and reasoning and their relation to natural language (see, e.g. [16, 31, 36]), a field where logical formats meet with statistical models of gradable beliefs.

When we talk about formalized semantic descriptions we should nevertheless acknowledge the role natural language implicitly plays as a description (or denotation) language. The implicit semantics human understanders attribute to the underlying representation *Playing_tennis* ({*Jack, John*}$_\text{ACT}$) is immediately rooted in our natural language understanding capabilities of English, e.g. that *Playing_tennis* is an (English) event descriptor (name) for the real-world activity of playing tennis. This straightforward mapping vanishes when different natural languages come into play. We might, for instance, switch to French *Jouer_au_tennis*, Dutch *tennissen*, Portuguese *jogar tênis* or German *Tennis_spielen*. The illusive semantic transparency vanishes completely when we go one crucial step further and move on to an entirely artificial representation format such as $X17$ ({$p4, q8$}$_\text{T5FG}$). From the perspective of formal semantics, mappings from these artificial designators to names belonging to the name spaces of domains of discourse for specific natural languages have to be provided (e.g. $X17 \rightarrow Playing\_tennis$). Furthermore, the truth value of the whole assertion has to be determined relative to some logical model, e.g. *Playing_tennis* ({*Jack, John*}$_\text{ACT}$) is a TRUE statement based on grounding the two human players in the domain of discourse and finding them properly related by the relation *Playing_tennis*. Technically, in a logical calculus, '$X17$ ({$p4, q8$}$_\text{T5FG}$)' is nevertheless equivalent to '*Playing_tennis* ({*Jack, John*}$_\text{ACT}$)' (given appropriate mappings), but the latter expression is much more intelligible for humans than the first one.

Even in a statistical approach to semantic symbol analysis relations are crucial for representation. However, any statistical relation between entities is an associative, not a truth-functional one. The meaning of an associative relation, however, is hard to imagine for humans because the association just provides a degree of strength for the terms involved, e.g. to co-occur in sentences or other formal document sections. So a statistical relation is an *(n-1)*-ary expression where the lacking nth term is the semantics-providing predicate symbol.

While a formal logical approach to the semantics of a natural language leads to a theoretically sound embedding into the framework of model-theoretically based symbol-oriented semantics, one might argue whether such a symbolic approach truly captures our intuitions about the *real(istic)* distinction between playing tennis, squash, football, hockey or checkers. This problem cannot be solved within any symbol-based approach (such as the logical one) since it deals with symbols (and, for logics, truth values assigned to symbols), only, that need some sort of interpretation; when we leave the realm of logical specification this interpretation is almost always natural language-bound. As an alternative, one might consider symbol-grounding approaches which abstract away from symbolic interpretation problems and relate symbols to physical parameters and perceptual data in the real world (e.g. spatio-temporal data points in a 3-D representation), which might then be susceptible to additional quantitative analysis (as outlined, e.g. by [14, 24] or [20]). In the long run, the symbol-grounding approach, with its direct ties to perceived reality, might link our human understanding of the world to quantitative categorical description systems mentioned in the introductory section. A serious problem for this approach is, however, to adequately deal with non-groundable terms which address non-physical entities (such as 'freedom', 'art' or 'knowledge').

2.3   Inference Rules: Formal Reasoning on Assertions

We will now consider methodological means to overcome the static representation of the world as sets of assertions (relations) and dynamically compute representation structures which render 'novel' knowledge. We, therefore, introduce *inference rules* as a formal device to enlarge the set of a priori given assertions (relations) by purely mechanical means (however, in the deductive reasoning mode, this knowledge is logically already entailed in the given set of assertions and rules). Consider the following, very simple example: 'All tennis playing events (represented by the predicate symbol '*Playing_tennis*') are also sports activities (represented by the predicate symbol '*Sports_activity*')' and 'All sports activities are also competition events (represented by the predicate symbol '*Competition_event*').' Accordingly, when we start with the initial knowledge base $KB_0 = \{Playing\_tennis\ (\{Jack,\ John\}_{ACT})\}$, by applying the above rules we end up with an enhanced knowledge base that incorporates two inferred predicates, $KB_{0*} = \{Playing\_tennis\ (\{Jack,\ John\}_{ACT}),$ $Sports\_activity\ (\{Jack,\ John\}_{ACT}),\ Competition\_event\ (\{Jack,\ John\}_{ACT})\}$. Note that we have not observed anything like a sports activity involving Jack and John, but we inferred the corresponding assertion '$Sports\_activity\ (\{Jack,\ John\}_{ACT})$' on the basis of the above-mentioned informal inference rule. Assuming a proper truth-preserving logical framework, we also know that if $Playing\_tennis(\{Jack,$ $John\}_{ACT})$ is TRUE, then $Sports\_activity\ (\{Jack,\ John\}_{ACT})$ will also be TRUE, as well as $Competition\_event\ (\{Jack,\ John\}_{ACT})$ will be TRUE. In a similar way, we may conclude that Jack's and John's outdoor tennis playing implies that the weather conditions during the match were dry assuming the informal inference rule: If $Playing\_tennis(\{Jack,\ John\}_{ACT,\ \_\_\_LOC})$ is TRUE, then $Weather\_Conditions$ $(dry,\ \_\_\_LOC)$ will also be TRUE (for outdoors tennis).

   A considerable part of our common-sense knowledge (as well as our scientific knowledge) is based on reasoning this way. Imagine a short story such as the following one: 'Jack played tennis with John. Jack's racket got damaged when he threw it on the ground and so he lost the match.' To grasp the plot of that story we have to know that a racket is the only device to play tennis. Tennis can only be successfully played with fully operational rackets. If a racket gets damaged in the course of the match and no proper substitute racket is available, the player with the damaged racket is most likely to lose the game because the player's ability to hit and smash the tennis ball properly is heavily impaired.

In the quantitative view of the world, a light-weight correspondence can be established between inference rules and algebraic composition rules, e.g. addition yields a new number (the sum) derived from a series of argument numbers. Whereas numerical addition works for some domains (e.g. when I add 5,000 € to my account which originally had a balance of 50,000 €, I end up with a new amount of 55,000 € on my account), for others it does not (e.g. addition is not defined for gene sequences or natural language utterances or speech signals as raw data input). Consequently, other operations on qualitative signature strings, e.g. splicing and copying (for molecular biology) or concatenating or parsing (for natural language processing), are required here.

3 Knowledge Resources for Semantic Computing: Terminologies, Thesauri and Ontologies

In the previous section, we have distinguished three major ingredients of semantic computing—terms and their internal, i.e. definitional, or external, i.e. relational, semantic structure, assertions (predicates) as a combination of terms which can formally be treated as relations, and inference rules as a formal means to mechanically generate additional assertions from a set of *a priori* given assertions where the extension of the original assertion set is licensed by the rules available.

We discussed these notions primarily with focus on how terms and assertions are structured. This issue is important because structural conventions are needed for computational processing. Once the knowledge *format* is agreed upon, another serious issue comes up for semantic computing, namely which resources are already available or still have to be supplied that capture concrete pieces of knowledge (*contents*) from different domains in the agreed upon format—common-sense knowledge as well as domain-specific scientific knowledge, e.g. from biology, medicine, physics, economics, arts, or agriculture.

Basically, there are currently two threads of development of such *knowledge resources* which constitute the foundations of semantic technologies—informal *terminologies* and (more) formal *ontologies*. Both capture the knowledge of some domain of discourse on the basis of specifying the relevant terms for such a domain and a restricted set of semantic relations linking these terms, mostly taxonomically ones. Yet, terms and relations often, though not always, lack a clearly defined formal meaning and thus appeal to human understanding of the terms' and relations' names or their verbal definitions (giving rise to the problems just discussed when natural language is used as a meta language for semantic description).

The life sciences are one of the most active scientific areas in which we observe a long-standing tradition in providing, actively maintaining and newly developing such knowledge repositories. The most important and freely available ones are organized in large umbrella systems. For instance, the *Unified Medical Language System* (UMLS)[12] assembles more than 100 medical source terminologies ranging from anatomy, pathology to diagnostic and clinical procedures. In a similar vein, the *Open Biological Ontologies* (OBO)[13] and the NCBO BIOPORTAL[14] constitute a collection of more than 80 and 200, respectively, biological ontologies, again covering many of the specialties in this field (e.g. genes and proteins, experimental methods, species, and tissues). These terminologies and ontologies vary tremendously in size (some few hold more than 70,000 base terms and 150,000 relations, others are small and cover no more than 1,000 terms and relations). Though being valuable resources on its own, semantic integration is usually out of reach because terminologies/ontologies being part of one of these umbrellas are not really linked to each other. This issue is dealt with in the field of *ontology alignment* [17], with its emphasis on finding lexical and structural indicators for matching and connecting different, yet semantically overlapping, knowledge resources in a larger and coherent knowledge assembly.

Whereas the life sciences are a particularly impressive and active application area for terminologies and ontologies, other areas, such as agriculture,[15] law,[16] or arts and architecture,[17] are also involved in such initiatives, though less committed. Finally, the requirements for dealing with everyday common-sense knowledge have also led to the emergence of general-purpose ontologies such as SENSUS[18] and CYC[19]. There are also borderline cases where lexical databases such as WORDNET[20] and FRAMENET[21] (both for the general English language) are considered as ontologies as well because they make heavy use of (verbally defined) semantic relations as part of their specifications. Two problems are common to all these approaches. First, how to cover a really large proportion of subject- as well as domain-specific terms and relations from a chosen domain, second, what types of semantic relations should be encoded?

---

[12]http://www.nlm.nih.gov/research/umls/

[13]http://www.obofoundry.org/

[14]http://bioportal.bioontology.org/

[15]http://agclass.nal.usda.gov/

[16]http://www.loc.gov/lexico/servlet/lexico/liv/brsearch.html?usr=pub-13:0&op=frames&db=GLIN

[17]http://www.getty.edu/research/tools/vocabularies/aat/index.html

[18]http://www.isi.edu/natural-language/resources/sensus.html

[19]http://www.cyc.com/

[20]http://wordnet.princeton.edu/

[21]https://framenet.icsi.berkeley.edu/fndrupal/

As to the first question, virtually thousands of person years have already been spent world-wide on the manual creation and maintenance of terminological and ontological resources. The results though being impressive in terms of the sheer size of some repositories are often controversial within the respective domains, when it comes to the underlying principles of the organization of terms or specific content areas, i.e. particular branches of a terminology. This is, to some degree, due to vague and often even lacking formal semantic criteria what a particular relation really should mean (see, e.g. [30, 43, 44], and also the remarks below). Note that any form of terminological specification which rests on natural language inherits, by default, all known problems of natural languages as specification languages (being inherently vague, imprecise, ambiguous, etc.). Attempts at automating the construction process of lexicons, terminologies or ontologies are still in their infancy (this area of research is called *lexical/semantic acquisition* (see, e.g. [8]) or *ontology learning* (see, e.g. [12]) and do currently not deliver reasonable and sound results, on a larger scale, at least.

As to the second question, concerning the types of semantic relations to be taken into consideration for terminology / ontology building, several disciplines contributed to such efforts. In structural linguistic semantics, as well as in the broad field of documentation languages (including, e.g. nomenclatures, terminologies, thesauri, classifications for scientific domains), which is part of information (or library) science, we witness a tradition of dealing with terms (in linguistic jargon, lemmata or lexical entries) and informal semantic relations among terms. These activities have mostly been focusing on the specification of synonyms, taxonomic and partonomic relations. Taxonomic and partonomic relations, e.g. *Heart* Is-a *Organ* and *Heart* Has-part *Heart Chamber*, cover terminological relationships but do not specify any empirical or propositional ones as, e.g. '*Tennis_playing*' or '*Is-effective-for*' do (when combined with proper arguments). Only in the life sciences domain, such propositional relations play an admittedly marginal, yet currently growing role (most ontologies lack them, for those which incorporate them their proportion is usually less than 5–10% of all relations available; around 80% deal with taxonomic and 10–15% with partonomic relations (as of the end of 2011)).

These semantic relations are assigned *a priori* to different terms by linguists, information scientists, terminologists, ontologists, or domain experts. There is also usually no serious attempt at specifying the regularities holding between hyponyms and hypernyms or partonyms and holonyms at a more general and formal level, at least in terminologies. Work on ontologies has made some progress in this field of research. One stream of activities deals with the specification of algebraic properties between terms linked by various semantic relations. For instance, the *Is-a* relation is assigned the properties of being transitive, antisymmetric, and reflexive (for further relations which also obey algebraic properties, see [42]).

Another way of dealing with the (formal) semantics of the *Is-a* relation is by defining necessary and sufficient criteria for definitional properties of terms and *computing* (rather than defining) relationships on the basis of these specifications. This brings us back to the Aristotalian approach to term definition where a *genus proximum* (the hypernym) is complemented by a set of *differentia specifica* (definitional properties which constrain the *genus proximum* to the subclass to be defined). Consider, e.g. the following two abstract term definitions, $term_1$: [p_1, p_2 and p_3] and $term_2$: [p_1, p_2, p_3 and p_4]. A formal reasoning engine (a terminological classifier) which is capable of interpreting the above property inclusion ($\{p_1, p_2$ and $p_3\}$ is a proper subset of $\{p_1, p_2, p_3$ and $p_4\}$) computes that $term_2$ is a hyponym of $term_1$, the hypernym, on the basis of formal property set inclusion, a principle called *term subsumption* [4]. An entire branch of restricted predicate logics, called *description logics*, deal with these kinds of terminological computations [2, 3].

Whereas the distinction between terminologies and ontologies remains fuzzy in the literature, we here state that *terminologies* are collections of terms which are either connected by semantic relations whose semantics is specified, if at all, in terms of natural language definitions that lack any or sufficient formal grounding, or provide natural language definitions of their terms (see, e.g. the glosses provided in WORDNET). On the other hand, *ontologies* are based on comprehensive formal conventions (e.g. algebraic properties of relations or subsumption criteria) and assorted reasoning devices (terminological classifiers) that can make use of term subsumption or other conceptual computation schemes relying, e g. upon algebraic properties of semantic relations.

If the Semantic Web builds on such semantic resources, who actually provides semantic meta data for raw data (i.e. *annotations*) using them? Currently, two approaches are pursued. The first one relies on human volunteers who painfully annotate small portions of Semantic Web pages by hand—sometimes relying on canonical semantic resources as those mentioned above, sometimes just using *ad hoc* non-normalized natural language terms and phrases for annotation purposes. The second approach to annotation makes use of computational means to render such semantic meta data automatically.

Obviously, this brings us back to the challenges of the computational analysis of natural languages. And, indeed, there is much prospect to supply suitable natural language processing methodologies and techniques for automatically structuring originally unstructured raw data (such as texts or speech). This field may thus become one of the core semantic technologies for the Semantic Web, complementing and at the same time reusing the whole area of knowledge representation, reasoning and ontology engineering. Automatic natural language processing (see, e.g. [33] or [28] for surveys) disposes of methodologies, particularly in its application areas, such as information extraction, text mining, automatic summarization and question answering, which are robust enough to operate under Web dimensions, scalable to the processing of millions of documents and capable of annotating fundamental semantic entities (based on named entity recognition devices) and their interrelations (using relation extraction engines) in the Web. Thus, this field might deliver technologies that will either replace the human in the annotation loop or provide computer-supported guidance for human annotation decisions by pre-selection of relevant documents and making annotation proposals which have to be acknowledged (or rejected) by human annotation supervisors.

4 Formalisms and Standards for Semantic Computing: RDF, RDF-S, and OWL

In the previous section, we focused primarily on the structure of terms and relations and how these constructs are used in terminologies and ontologies. We now turn to computational issues related to them—representation conventions and reasoning engines using these shared formats. We already introduced inference rules as a formal means to mechanically generate additional assertions from a set of *a priori* given assertions where the extension of the original assertion set is licensed by rules (truth-preserving ones when logical calculi are applied deductively). Formally, these deductive extensions are already logically implied by the initial assertion set together with the rules being supplied and, thus, only for human perception constitute 'new' knowledge.

These mechanics, their formal underpinnings and implications are well known in the AI knowledge representation community for quite a while. However, harmful formal properties of reasonably powerful logical calculi (e.g. various forms of description logics (DL), first-order predicate logics (FOPL), as well as even more sophisticated, e.g. modal, logics) concerning their computational tractability (e.g. lacking decidability of FOPL, NP-completeness of subsumption in DL, etc.)[22] are a strong impediment to the application of these formal systems in successful real-world applications.

The Semantic Web as a data container adds further constraints to the usability of such formal systems. Prime among them is the requirement that supersized volumes of data have to be accounted for. Under these circumstances, completeness of (formal) specifications and their consistency can never be granted in the Web. Also the Semantic Web is constantly growing so that any fixed number of data or knowledge items (the Web considered as a static knowledge base) cannot be assumed. Data snapshots are outdated in the very moment they are made. Not to mention the fact that Web data are notoriously 'dirty', i.e. uncertain, incomplete, vague, imprecise, contradictory, etc. and contributions vary significantly in terms of quality, trustability and objectiveness. Hence, rather than pursuing a 'strong' approach to formal knowledge representation and reasoning requiring clean and, therefore, artificial toy world assumptions about the nature of the raw data, the emerging semantic technologies aim at a compromise between enhanced expressivity of the underlying representation structures and the operations that can be performed on them in the large, i.e. on a Web-scale basis, also being aware of 'hostile' degradations and deformations of these data.

The formal mechanisms which we have discussed in the previous sections constitute the core of the emerging semantic computing paradigm. Relational tuples with reference to some state of the world are known in linguistic semantics under the heading of predicate-argument structures for a long time (although formal reasoning

---

[22]http://www.cs.man.ac.uk/~ezolin/dl/

and the inferential derivation of new assertions has not been much of a concern in this field), whereas the roots of description logics (and thus the rich potential for terminological reasoning) can be traced in the history of AI back to the early eighties of the past century (in particular, their emergence from 'structured' knowledge representation formalisms, such as early semantic networks, frames or scripts). So the structural format and inferential power of knowledge representation schemes is not the ground-breaking innovation behind semantic computing given this legacy.

The reason why these methods recently became so important for semantic computing is the fact that *standards* were emerging and a community has formed that adheres to them in their daily work. Relational tuples, for instance, became a W3C standard under the name of *Resource Description Framework* (RDF).[23] RDF is based on the idea of identifying 'things' using Web identifiers (called *Uniform Resource Identifiers* (URIs)), and describing resources in terms of simple properties and property values. The convention basically requires statements about the real world to be made in the format of tuples which are composed of three basic entities, namely *(subject—predicate—object)*. The 'thing' a statement is about is called the *subject*. The part that identifies the property or characteristic of the subject that a statement specifies is called the *predicate*, and the part that identifies the value of that property is called the *object*. For instance, if a photograph under http://www. PicofJohn.jpg (the subject) depicts (the predicate) 'John' as a tennis player (the object), then we might capture this information as an RDF statement in the following way: (http://www.PicofJohn.jpg,http://purl.org/dc/elements/1.1/depiction, 'John as a tennis player') Huge amounts of such triples are then collected in so-called RDF or Large Triple Stores which constitute, in essence, a structured database of assertions about the world.[24]

RDF is quite limited in terms of expressivity and can be coupled with the *Simple Knowledge Organization System* (SKOS)[25] for lightweight knowledge sharing and linking on the Web (e.g. using informal thesauri and classifications). An extension of RDF, RDF-SCHEMA (RDFS) or *RDF Vocabulary Description Language*,[26] allows to state semantic relations among the terms used in RDF statements. Semantically even more powerful, however, is to link RDF structures with inference rules [27] such that an embedding of RDF into OWL is achieved. OWL corresponds to a family of description logics which incorporate different expression layers, ranging from full OWL via OWL-DL to OWL-LITE. While expressivity decreases from full to lite variants, computational properties get more handsome (e.g. full OWL or OWL with the *Semantic Web Rule Language* (SWRL) are undecidable, whereas OWL-DL and OWL-LITE are decidable; see, e.g. [34]). OWL, unlike RDF, still has not become an industry standard up until now, but has rather become a *de facto* standard for the Semantic Web community.

---

[23]http://www.w3.org/RDF/

[24]http://www.w3.org/wiki/LargeTripleStores

[25]http://www.w3.org/2001/sw/wiki/SKOS

[26]http://www.w3.org/TR/rdf-schema/

URI, RDF, RDFS, SKOS and OWL are computable representation schemes. This is fine from a methodological perspective but unless massive amounts of contents are encoded and publicly made accessible in this format they are not much of a help for large-scale semantic computing. Fortunately, the vast majority of ontologies developed these days are not only specified in RDF format but, even more so, are made available in some form of OWL. Hence, large amounts of not only common-sense knowledge representation structures (SENSUS, CYC) but also domain-specific, special-purpose knowledge (e.g. for medicine (FMA,[27] NCI,[28] etc.) or biology (GO,[29] RO,[30] etc.)) can be reused for semantic computation.

5   Semantic Technologies: The Paradigm and the Reality

The semantics of qualitative knowledge of the world have long been disregarded by computer science and industry. This is beginning to change these days as computationally adequate representation formats (RDF, OWL, etc.) and corresponding software (e.g. RDF databases, OWL classifiers) are being made available and robust enough for large-scale knowledge processing. While the prospects look very positive for semantic technologies, the field will have to solve several crucial problems some of which are outlined below.

1. Currently, a large diversity and heterogeneity characterizes the wealth of available and emerging ontologies. Lacking though are robust mechanisms which interlink single ontologies by means of automatic or computer-supported interactive *ontology alignment*. The problems encountered are: different perspectives (views) in different ontologies on the same subject area (e.g. should a disease ontology assume ideal organism functions for reference purposes or assume pathologies, and if so, which ones, as a comparison standard?), different levels of granularity of knowledge (shallow (layman) *versus* detailed (expert) knowledge) and different levels of definitional rigidity and empirical correctness—to mention just a few hurdles for the *integration* and *interoperability* of collections of ontologies.

2. Many domains of discourse lack ontological descriptions or, if available, suffer from limited depth and coverage of their specifications. Since building and maintaining semantic descriptions is extremely time-consuming and intellectually challenging, *automatic knowledge acquisition* and *ontology learning* have become premium research topics. The grand idea is to use raw Web data (free text, definitional statements, e.g. from WIKIPEDIA, perhaps complemented

---

[27]http://trac.biostr.washington.edu/trac/wiki/FMAInOWL

[28]http://ncicb.nci.nih.gov/download/evsportal.jsp

[29]http://www.geneontology.org/GO.format.shtm

[30]http://obofoundry.org/ro/

by clarification dialogs between a computational learner and a human expert) and apply machine learning and natural language techniques for this structure-building process.

3. One might then go one step further and provide computational solutions for the *data annotation* problem. Given the obvious need for more complete semantic meta data descriptions in terms of ontologies, natural language processing techniques might be applied which automatically generate massive amounts of semantic meta data for raw data on the fly.

4. The semantic encoding, be it in terms of RDF tuples or in terms of OWL-LITE, is far too flat, i.e. inexpressive, to capture the richness of expressivity of natural languages, in general. The mismatch is hard to balance, in principle. Increasing the *expressivity* of formal semantic calculi inevitably increases the level of *computational complexity* up to a point where these formal systems rapidly become computationally intractable. There is no clear way out of this dilemma since proven mathematical findings put absolute limits on the solution space for semantic computability.

   Despite many formal barriers for highly expressive semantic computing, we might still find reasonable processing *heuristics*, e.g. by prioritizing different computational alternatives along the line of Markov Logics [16], perhaps, including cost and utility considerations as part of abductive reasoning schemes [7, 26] or incorporating anytime algorithms [23]. Another solution might be to allocate cheap computations to bulk processing, and reserve 'expensive' computational schemes only for presumably extra-rewarding, smaller scaled semantic processing tasks (e.g. along the lines of the theory of granularity [25] or abstractions [19]).

5. A link between quantitative and qualitative knowledge representation schemes is also missing (e.g. moving from the computation of a Fourier transform or the eigenvalue of a matrix to deductive conceptual reasoning within the framework of description logics; see, e.g. [32]). Obvious application areas are the wide range of life sciences, numerous areas of engineering and the natural sciences, etc. It might well be that such efforts will lead to a natural way of grounding symbols in metrical, quantitative representation schemes.

6. *Trustability* of data and services in the Web has become a major issue in the light of the ever-increasing importance of Web-based knowledge. With the marriage of current social computation models in the field of recommender systems [37], the decoding of subjective (i.e. evaluative, opinion-bearing) language by natural language processing techniques [46], and cryptographic security models [21], the detection of cheated, unsafe, or even criminal contents might become feasible. Note that the trust level is already an integral part of the W3C layer model.[31] In the end, efforts along these dimensions will enhance semantic, content-oriented computation up to the level of *pragmatic computing* where the usage and usability of content in social settings becomes the crucial issue for computation.

---

[31]http://www.w3.org/wiki/SemanticWebArchitecture

7. Today's computation problems are embedded in complex decision support and knowledge management scenarios. Knowledge representation, and ontologies as their domain-specific offspring, is but one methodology needed here. Others are search, learning, planning, constraint satisfaction and scheduling procedures. Therefore, stand-alone systems for knowledge representation, reasoning and content analytics might only be of limited value under such complex task scenarios. Hence, a distributed problem solving architecture as part of smart workflows is needed which incorporate the semantics of the domain and the services to be supported. In such a framework, the knowledge proper plays a crucial though not the only role and will thus be complemented by models of the tasks to be achieved and the environment in which the operations for task fulfillment take place.

Semantic technologies are the methodological backbone for the evolving Semantic Web. They thus complement current efforts which aim at the seamless availability of heterogeneous data (*Linked Open Data*)[32] and collaborative efforts for semantic annotation of raw data captured in metaphors such as the 'wisdom of the crowds' (see, e.g. [29]), or businesses such as the *Mechanical Turk*.[33] The success of semantic technologies is both due to emerging large-scale industrial efforts devoted to ontology engineering, knowledge representation and formal reasoning as well as the increasing analytic power and robust scalability of natural language technology systems. The first stream of work provides language-independent semantic knowledge structures, the second stream services the needs for representation standards and assorted reasoning technology, whereas the third one deals with a prime carrier of qualitative knowledge, namely natural languages.

Given this dependence on natural languages informally encoded semantic structures have to be transformed into formally coded and efficiently accessible and processable ones. Since the proportion of verbally encoded knowledge is much larger than that of rigorously defined quantitative knowledge and since this verbalized knowledge is recognized as being extremely important for further economic and scientific progress, we might just witness the beginning of a new computational era that substantially enhances, perhaps even fundamentally changes, the established notion of 'computation'.

References

1. Altschul, S.F., Gish, W., Miller, W., Myers, E.W., Lipman, D.J.: Basic local alignment search tool. J. Mol. Biol. **215**(3), 403–410 (1990)
2. Baader, F., Calvanese, D., McGuinness, D.L., Nardi, D., Patel-Schneider, P.F. (eds.): The Description Logic Handbook: Theory, Implementation, and Applications, 2nd edn. Cambridge University Press, Cambridge (2007)

---

[32]http://www.w3.org/standards/semanticweb/data

[33]https://www.mturk.com/mturk/welcome

3. Baader, F., Horrocks, I., Sattler, U.: Description logics. In: van Harmelen, F., Lifschitz, V., Porter, B. (eds.) Handbook of Knowledge Representation, pp. 135–180. Elsevier, Amsterdam (2008)

4. Baader, F., Nutt, W.: Basic description logics. In: Baader, F., Calvanese, D., McGuinness, D.L., Nardi, D., Patel-Schneider, P.F. (eds.) The Description Logic Handbook: Theory, Implementation, and Applications, 2nd edn., pp. 47–100. Cambridge University Press, Cambridge (2007)

5. Berners-Lee, T., Hendler, J., Lassila, O.: The Semantic Web: a new form of Web content that is meaningful to computers will unleash a revolution of new possibilities. Scientif. Am. **284**(5), 34–43 (2001)

6. van Benthem, J., Ter Meulen, A.: Handbook of Logic and Language, 2nd edn. Elsevier, Amsterdam (2010)

7. Blythe, J., Hobbs, J.R., Domingos, P., Kate, R., Mooney, R.: Implementing weighted abduction in Markov Logic. In: IWCS 2011 — Proceedings of the 9th International Conference on Computational Semantics, pp. 55–64 (2011)

8. Boguraev, B., Pustejovsky, J. (eds.): Corpus Processing for Lexical Acquisition. MIT, New York (1996)

9. Brachman, R.J.: What *Is-A* is and isn't: an analysis of taxonomic links in semantic networks. IEEE Comput. **16**(10), 30–36 (1983)

10. Brachman, R., Levesque, H.: Knowledge Representation and Reasoning. Morgan Kaufmann, Los Altos (2004)

11. Chierchia, G., McConnell-Ginet, S.: Meaning and Grammar: An Introduction to Semantics, 2nd edn. MIT, Cambridge, MA (2000)

12. Cimiano, P.: Ontology Learning and Population from Text. Springer, Berlin (2006)

13. Clocksin, W.F., Mellish, C.S.: Programming in Prolog. Springer, Berlin (2003)

14. Cottrell, G.W., Bartell, B., Haupt, C.: Grounding meaning in perception. In: GWAI-90 — Proceedings of the 14th German Workshop on Artificial Intelligence, pp. 307–321 (1990)

15. Cruse, D.A.: Lexical Semantics. Cambridge University Press, Cambridge (1986)

16. Domingos, P., Lowd, D.: Markov Logic: An Interface Layer for Artificial Intelligence. Morgan & Claypool, San Rafael (2009)

17. Euzenat, J., Shvaiko, P.: Ontology Matching. Springer, Berlin (2007)

18. Fox, C., Lappin, S.: An expressive first-order logic with flexible typing for natural language semantics. Logic J. Int. Group Pure Appl. Logic **12**, 135–168 (2004)

19. Giunchiglia, F., Walsh, T.: A theory of abstraction. Artif. Intell. **57**(2–3), 323–389 (1992)

20. Glenberg, A.M., Robertson, D.A.: Symbol grounding and meaning: a comparison of high-dimensional and embodied theories of meaning. J. Memory Lang. **43**(3), 379–401 (2000)

21. Goldreich, O.: Foundations of Cryptography. Volume 1: Basic Tools. Oxford University Press, Oxford (2007)

22. Halevy, A., Norvig, P., Pereira, F.: The unreasonable effectiveness of data. IEEE Intell. Syst. **24**(2), 8–12 (2009)

23. Hansen, E.A., Zilberstein, S.: Monitoring and control of anytime algorithms: a dynamic programming approach. Artif. Intell. **126**(1–2), 139–157 (2001)

24. Harnad, S.: The symbol grounding problem. Physica D **42**(1–3), 335–346 (1990)

25. Hobbs, J.R.: Granularity. In: IJCAI-85 — Proceedings of the 9th International Joint Conference on Artificial Intelligence, pp. 432–435 (1985)

26. Hobbs, J.R., Stickel, M.E., Appelt, D.E., Martin, P.A.: Interpretation as abduction. Artif. Intell. **63**(1–2), 69–142 (1993)

27. Horrocks, I., Patel-Schneider, P.F., van Harmelen, F.: From SHIQ and RDF to OWL: the making of a Web Ontology Language. J. Web Semant. **1**(1), 7–26 (2003)

28. Jurafsky, D., Martin, J.H.: Speech and Language Processing. An Introduction to Natural Language Processing, Computational Linguistics, and Speech Recognition, 2nd edn. Prentice Hall, Englewood Cliffs (2009)

29. Kittur, A., Kraut, R.E.: Harnessing the wisdom of the crowds in WIKIPEDIA: quality through coordination. In: CSCW '08 — Proceedings of the 2008 ACM Conference on Computer Supported Cooperative Work, pp. 37–46 (2008)

30. Köhler, J., Munn, K., Rüegg, A., Skusa, A., Smith, B.: Quality control for terms and definitions in ontologies and taxonomies. BMC Bioinformat. **7**, 212 (2006)

31. Koller, D., Friedman, N.: Probabilistic Graphical Models: Principles and Techniques. MIT, New York (2009)

32. Lutz, C.: Adding numbers to the SHIQ description logic: first results. In: KR-02 — Proceedings of the 8th International Conference on Principles and Knowledge Representation and Reasoning, pp. 191–202 (2002)

33. Manning, C., Schütze, H.: Foundations of Statistical Natural Language Processing. MIT, New York (1999)

34. Motik, B.: On the properties of metamodeling of OWL. J. Logic Comput. **17**(4), 617–637 (2007)

35. Nilsson, N.: Principles of Artificial Intelligence. Morgan Kaufmann, Los Altos (1980)

36. Pearl, J.: Probabilistic Reasoning in Intelligent Systems: Networks of Plausible Inference. Morgan Kaufmann, Los Altos (1988)

37. Ricci, F., Rokach, L., Shapira, B.: Recommender Systems Handbook. Springer, Berlin (2011)

38. Russell, S.J., Norvig, P.: Artificial Intelligence: A Modern Approach, 3rd edn. Prentice Hall, Englewood Cliffs (2010)

39. Schulz, S., Kumar, A., Bittner, T.: Biomedical ontologies: what *part-of* is and isn't. J. Biomed. Informat. **39**(3), 350–361 (2006)

40. Searls, D.B.: The language of genes. Nature **420**(6912), 211–217 (2002)

41. Shariff, A.A., Hussain, M.A., Kumar, S.: Leveraging unstructured data into intelligent information: analysis & evaluation. In: Proceedings of the 2011 International Conference on Information and Network Technology, pp. 153–157 (2011)

42. Smith, B., Ceusters, W., Klagges, B., Köhler, J., Kumar, A., Lomax, J., Mungall, C., Neuhaus, F., Rector, A.L., Rosse, C.: Relations in biomedical ontologies. Genome Biol. **6**, R46 (2005)

43. Smith, B., Köhler, J., Kumar, A.: On the application of formal principles to life science data: a case study in the *Gene Ontology*. In: DILS 2004—Data Integration in the Life Sciences, pp. 79–94 (2004)

44. Smith, B., Rosse, C.: The role of foundational relations in the alignment of biological ontologies. In: MedInfo 2004—Proceedings of the 11th World Congress on Medical Informatics, pp. 444–448 (2004)

45. Sowa, J.F.: Conceptual Structures: Information Processing in Mind and Machine. Addison-Wesley, Reading (1984)

46. Wiebe, J., Wilson, T., Bruce, R., Bell, M., Martin, M.: Learning subjective language. Comput. Linguist. **30**(3), 277–308 (2004)

What Are Ontologies Good For?

Ian Horrocks

Abstract Ontology, in its original philosophical sense, is a fundamental branch of metaphysics focusing on the study of existence; its objective is to determine what entities and types of entities actually exist and thus to study the structure of the world. In contrast, in computer science an ontology is an engineering artifact, usually a "conceptual model" of (some aspect of) the world, typically formalised as a logical theory. Formalising an ontology using a suitable logic opens up the possibility of using automated reasoning to support both ontology design and deployment. The value of such support has already been demonstrated in medical applications, where it has been used to help repair and enrich ontologies that play an important role in patient care.

Even with the aid of reasoning enabled tools, developing and maintaining good quality ontologies is a difficult and costly task, and problems related to the availability of good quality ontologies threaten to limit the deployment of ontology-based information systems. This has resulted in ontology engineers increasingly looking to the philosophy community for possible solutions, and in particular as a source of relevant expertise in the organisation and formalisation of knowledge.

1 Introduction

Ontology, in its original philosophical sense, is a fundamental branch of metaphysics focusing on the study of existence; its objective is to determine what entities and types of entities actually exist and thus to study the structure of the world. The study of ontology can be traced back to the work of Plato and Aristotle and includes the development of hierarchical categorisations of different kinds of entity and the

I. Horrocks (✉)
Oxford University, Department of Computer Science, Oxford, UK
e-mail: ian.horrocks@cs.ox.ac.uk

B.-O. Küppers et al. (eds.), *Evolution of Semantic Systems*,
DOI 10.1007/978-3-642-34997-3_9, © Springer-Verlag Berlin Heidelberg 2013

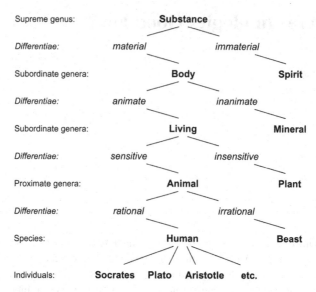

Fig. 1 Tree of Porphyry

features that distinguish them: the well-known "tree of Porphyry", for example, identifies animals and plants as sub-categories of living things distinguished by animals being *sensitive* and plants being *insensitive* (see Fig. 1).

In contrast, in computer science an ontology is an engineering artifact, usually a so-called "conceptual model" of (some aspect of) the world; it introduces vocabulary describing various aspects of the domain being modelled and provides an explicit specification of the intended meaning of the vocabulary by describing the relationships between different vocabulary terms. These relationships do, however, invariably include classification-based relationships not unlike those used in Porphyry's tree.

In a logic-based (formal) ontology, the vocabulary can be thought of as predicates and constants. For example, Animal and Plant can be thought of as (unary) predicates and Socrates, Plato and Aristotle as constants. Relationships between vocabulary terms can be specified using *axioms*, logical sentences such as

$$\forall x [\mathsf{Human}(x) \rightarrow \mathsf{Animal}(x) \wedge \mathsf{Sensitive}(x)]$$

$$\mathsf{Human}(\mathsf{Socrates})$$

An ontology can be thought of simply as a set of such axioms—i.e., a logical theory.

Viewing an ontology as a logical theory opens up the possibility of using automated reasoning to check, e.g., internal consistency (the theory does not entail "false") and other (non-) entailments. This is not just a theoretical possibility, but is the raison d'être of many logic-based ontology languages, in particular those based on description logics—for these languages, sophisticated reasoning tools have been developed and are in widespread use. Such tools can be used, for example, to check for *concept subsumption*, i.e., an entailment of the form $\mathcal{O} \models \forall x[C(x) \rightarrow D(x)]$ for an ontology $\mathcal{O}$ and concepts C and D.

The (possible) existence of such tools was an important factor in the design of the OWL ontology language [13] and its basis in description logics. OWL was initially developed for use in the so-called Semantic Web; the availability of description logic-based reasoning tools has, however, contributed to the increasingly widespread use of OWL, not only in the Semantic Web per se, but as a popular language for ontology development in fields as diverse as biology [22], medicine [5], geography [6], geology [27], astronomy [3], agriculture [24], and defence [18]. Applications of OWL are particularly prevalent in the life sciences, where it has been used by the developers of several large biomedical ontologies, including the Biological Pathways Exchange (BioPAX) ontology [21], the GALEN ontology [20], the Foundational Model of Anatomy (FMA) [5], and the National Cancer Institute thesaurus [10].

The importance of reasoning support was highlighted by [16], which described a project in which the Medical Entities Dictionary (MED), a large ontology used at the Columbia Presbyterian Medical Center, was checked using a reasoner. This check revealed "systematic modelling errors", and many missed concept subsumptions, the combination of which "could have cost the hospital many missing results in various decision support and infection control systems that routinely use MED to screen patients".

2 The Web Ontology Language OWL

The Web Ontology Language (OWL) [19] is currently by far the most widely used ontology language. OWL was developed by a World Wide Web Consortium (W3C) working group in order to extend the capabilities of the Resource Description Framework (RDF), a language for representing basic information about entities and relationships between them [13]. OWL exploited existing work on languages such as OIL [4] and DAML + OIL [12] and, like them, was based on a description logic (DL).

Description logics (DLs) are a family of logic-based knowledge representation formalisms; they are descendants of Semantic Networks [29] and KL-ONE [2]. These formalisms all adopt an object-oriented model, similar to the one used by Plato and Aristotle, in which the domain is described in terms of individuals, *concepts* (usually called *classes* in ontology languages), and *roles* (usually called *relationships* or *properties* in ontology languages). Individuals, e.g., "Socrates", are the basic elements of the domain; concepts, e.g., "Human", describe sets of individuals having similar characteristics; and roles, e.g., "hasPupil" describe relationships between pairs of individuals, such as "Socrates hasPupil Plato".

As well as *atomic* concept names such as Human, DLs also allow for concept descriptions to be composed from atomic concepts and roles. Moreover, it is possible to assert that one concept (or concept description) is subsumed by (is a sub-concept of), or is exactly equivalent to, another. This allows for easy extension of the vocabulary by introducing new names as abbreviations for descriptions. For example, using standard DL notation, we might write:

$$\text{HappyParent} \equiv \text{Parent} \sqcap \forall \text{hasChild}.(\text{Intelligent} \sqcup \text{Athletic})$$

This introduces the concept name HappyParent and asserts that its instances are just those individuals that are instances of Parent, and all of whose children are instances of either Intelligent or Athletic.

Another distinguishing feature of DLs is that they are logics and so have a formal semantics. DLs can, in fact, be seen as decidable subsets of first-order predicate logic, with individuals being equivalent to constants, (atomic) concepts to unary predicates and (atomic) roles to binary predicates. Similarly, complex concepts are equivalent to formulae with one free variable, and standard axioms are equivalent to (bi-) implications with the free variable universally quantified at the outer level. For example, the concept used above to describe a happy parent is equivalent to the following formula

$$\text{Parent}(x) \wedge \forall y[\text{hasChild}(x, y) \rightarrow (\text{Intelligent}(y) \vee \text{Athletic}(y))]$$

and the DL axiom introducing the concept name HappyParent is equivalent to the following bi-implication:

$$\forall x[\text{HappyParent}(x) \Longleftrightarrow \text{Parent}(x) \wedge \forall y[\text{hasChild}(x, y) \rightarrow (\text{Intelligent}(y) \vee \text{Athletic}(y))]]$$

As well as giving a precise and unambiguous meaning to descriptions of the domain, the use of a logic, and in particular of a decidable logic, also allows for the development of reasoning algorithms that can be used to answer complex questions about the domain. An important aspect of DL research has been the design of such algorithms and their implementation in (highly optimised) reasoning systems that can be used by applications to help them "understand" the knowledge captured in a DL-based ontology.

A given DL is characterised by the set of constructors provided for building concept descriptions. These typically include at least intersection ($\sqcap$), union ($\sqcup$) and complement ($\neg$), as well as restricted forms of existential ($\exists$) and universal ($\forall$) quantification, which in OWL are called, respectively, *someValuesFrom*, and *allValuesFrom* restrictions. OWL is based on a very expressive DL called $\mathcal{SHOIN}$ that also provides cardinality restrictions ($\geq$, $\leq$) and enumerated classes (called *oneOf* in OWL) [11, 13]. Cardinality restrictions allow, e.g., for the description of a concept such as people who have at least two children, while enumerated classes allow for classes to be described by simply enumerating their instance, e.g.,

$$\text{EUcountries} \equiv \{\text{Austria}, \ldots, \text{UK}\}$$

Fig. 2 OWL constructors

Constructor	DL Syntax	Example
intersectionOf	$C_1 \sqcap \ldots \sqcap C_n$	Human $\sqcap$ Male
unionOf	$C_1 \sqcup \ldots \sqcup C_n$	Doctor $\sqcup$ Lawyer
complementOf	$\neg C$	$\neg$Male
oneOf	$\{x_1 \ldots x_n\}$	{john, mary}
allValuesFrom	$\forall P.C$	$\forall$hasChild.Doctor
someValuesFrom	$\exists r.C$	$\exists$hasChild.Lawyer
hasValue	$\exists r.\{x\}$	$\exists$citizenOf.{USA}
minCardinality	$(\geqslant n\ r)$	$(\geqslant 2$ hasChild)
maxCardinality	$(\leqslant n\ r)$	$(\leqslant 1$ hasChild)
inverseOf	r^-	hasChild$^-$

Fig. 3 An example of OWL's RDF syntax

```
<owl:Class>
  <owl:intersectionOf rdf:parseType=" collection">
    <owl:Class rdf:about="#Parent"/>
    <owl:Restriction>
      <owl:onProperty rdf:resource="#hasChild"/>
      <owl:allValuesFrom>
        <owl:unionOf rdf:parseType=" collection">
          <owl:Class rdf:about="#Intelligent"/>
          <owl:Class rdf:about="#Athletic"/>
        </owl:unionOf>
      </owl:allValuesFrom>
    </owl:Restriction>
  </owl:intersectionOf>
</owl:Class>
```

$\mathcal{SHOIN}$ also provides for transitive roles, allowing us to state, e.g., that if y is an ancestor of x and z is an ancestor of y, then z is also an ancestor of x, and for inverse roles, allowing us to state, e.g., that if z is an ancestor of x, then x is also an descendent of z. The constructors provided by OWL, and the equivalent DL syntax, are summarised in Fig. 2.

In DLs it is usual to separate the set of statements that establish the vocabulary to be used in describing the domain (what we might think of as the schema) from the set of statements that describe some particular situation that instantiates the schema (what we might think of as data); the former is called the TBox (Terminology Box) and the latter the ABox (Assertion Box). An OWL ontology is simply equivalent to a set of $\mathcal{SHOIN}$ TBox and ABox statements. This mixing of schema and data is quite unusual (in fact ontologies are usually thought of as consisting only of the schema part), but does not affect the meaning—from a logical perspective, $\mathcal{SHOIN}$ KBs and OWL ontologies are just sets of axioms.

The main difference between OWL and $\mathcal{SHOIN}$ is that OWL ontologies use an RDF-based syntax intended to facilitate their use in the context of the Semantic Web. This syntax is rather verbose and not well suited for presentation to human beings. Figure 3, for example, illustrates how the description of happy parent given above would be written in OWL's RDF syntax.

3   Ontologies and Databases

Ontologies and ontology-based information systems are closely related in both form and function to databases and database systems: both support the development of domain models that can be queried and updated. Figure 4, for example, illustrates a database entity relationship (ER) schema that models (a fragment of) a university domain in much the same way as the conceptual (TBox) part of an ontology, and (some of) the information it captures could be expressed using the following DL axioms:

$$\text{Course} \sqsubseteq (\leq 1 \text{ Tby}) \sqcap (\geq 1 \text{ Tby}) \sqcap (\forall \text{Tby.Teacher}) \tag{1}$$

$$\exists \text{Tby}.\top \sqsubseteq \text{Course} \tag{2}$$

$$\text{GradStudent} \sqsubseteq \text{Student} \tag{3}$$

where (1) states that every course is taught by exactly one teacher, (2) states that only courses can be taught (i.e., being taught by something implies being a course), and (3) states that graduate students are a subclass of students.

Although there are clear analogies between databases and OWL ontologies, there are also important differences. Unlike databases, OWL has a so-called open world semantics in which missing information is treated as unknown rather than false, and OWL axioms behave like inference rules rather than database constraints. In the above axioms, for example, it is stated that only courses can be taught; in OWL, if we know that IntroductionToAI is taught by Dave, this leads to the implication that IntroductionToAI is a Course—if we were to query the ontology for instances of Course, then IntroductionToAI would be part of the answer. In a database setting the schema is interpreted as a set of constraints on the data: adding the fact that IntroductionToAI is taught by Dave without IntroductionToAI being already *known* to be a Course would lead to an invalid database state, and such an update would therefore be rejected by a database management system as a constraint violation.

In contrast to databases, OWL also makes no unique name assumption (UNA). For example, from axiom (1) we know that a course can be taught by only one teacher, so additionally asserting that IntroductionToAI is taught by David would lead to the implication that Dave and David are two names for the same individual. In a database setting this would again be treated as a constraint violation. Note that in OWL it is *possible* to assert (or infer) that two different names do *not* refer to the same individual; if such an assertion were made about Dave and David, then asserting that IntroductionToAI is taught by both Dave and David would make the ontology inconsistent. Unlike database management systems, ontology tools typically don't reject updates that result in the ontology becoming wholly or partly inconsistent, they simply provide a suitable warning.

The treatment of schema and constraints in a database setting means that they can be ignored at query time—in a valid database instance all the schema constraints must already be satisfied. This makes query answering very efficient: in order to

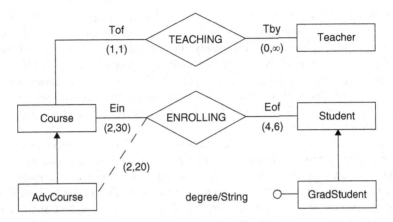

Fig. 4 An entity relationship schema

determine if IntroductionToAI is in the answer to a query for courses, it is sufficient to check if this fact is explicitly present in the database. In OWL, the schema plays a much more important role and is actively considered at query time. This can be very powerful and makes it possible to answer conceptual as well as extensional queries—for example, we can ask not only if Dave is a Teacher, but also if it is the case that anybody teaching a Course must be a Teacher. It does, however, make query answering *much* more difficult (at least in the worst case): in order to determine if Dave is in the answer to a query for teachers, it is necessary to check if Dave would be an instance of Teacher in every possible state of the world that is consistent with the axioms in the ontology. Query answering in OWL is thus analogous to theorem proving, and a query answer is often referred to as an entailment.

4 Ontology Reasoning

The design and implementation of reasoning systems is an important aspect of DL research and, as mentioned above, the availability of such reasoning systems was one of the motivations for basing OWL on a DL. This is because reasoning can be used in tools that support both the design of high quality ontologies and the deployment of ontologies in applications.

4.1 Reasoning at Design Time

Ontologies are often large and complex: the well-known SNOMED clinical terms ontology includes, for example, more than 400,000 class names [25]. Building and maintaining such ontologies is very costly and time consuming, and providing tools and services to support this "ontology engineering" process is of crucial importance to both the cost and the quality of the resulting ontology. State-of-the-art ontology development tools, such as SWOOP [14], Protégé 4 [17], and TopBraid Composer (see http://www.topbraidcomposer.com/), use a DL reasoner, such as FaCT++ [28], Racer [9] or Pellet [23], to provide feedback to the user about the logical implications of their design. This typically includes (at least) warnings about inconsistencies and redundancies.

An inconsistent (sometimes called unsatisfiable) class is one whose description is "over-constrained", with the result that it can never have any instances. This is typically an unintended feature of the design—why introduce a name for a class that can never have any instances—and may be due to subtle interactions between axioms. It is, therefore, very useful to be able to detect such classes and bring them to the attention of the ontology engineer. For example, during the development of an OWL ontology at the NASA Jet Propulsion Laboratory, the class "OceanCrustLayer" was found to be inconsistent. This was discovered (with the help of debugging tools) to be the result of its being defined to be both a region and a layer, one of which (layer) was a 2-dimensional object and the other a 3-dimensional object, where the axioms describing 2-dimensional and 3-dimensional objects ensured that these two classes were disjoint (had no instances in common). The inconsistency thus highlighted a fundamental error in the design of the ontology, discovering and repairing which obviously improved the quality of the ontology.

It is also possible that the descriptions in the ontology mean that two classes necessarily have exactly the same set of instances, i.e., that they are alternative names for the same class. This may be desirable in some situations, e.g., to capture the fact that "Myocardial infarction" and "Heart attack" mean the same thing. It could, however, also be the inadvertent result of interactions between descriptions, and so it is also useful to be able to alert users to the presence of such "synonyms". For example, when developing a medical terminology ontology a domain expert added the following two axioms:

$$\mathsf{AspirinTablet} \equiv \exists \mathsf{hasForm.Tablet}$$
$$\mathsf{AspirinTablet} \sqsubseteq \mathsf{AspirinDrug}$$

intending to capture the information that aspirin tablets are just those aspirin drugs that have the form of a tablet. Instead, these axioms had the effect of making *every* kind of tablet be an aspirin tablet. This was immediately corrected when the reasoner alerted the domain expert to the unexpected equivalence between Tablet and AsprinTablet.

In addition to checking for inconsistencies and synonyms, ontology development tools usually check for implicit subsumption relationships and update the class hierarchy accordingly. This is also a very useful design aid: it allows ontology developers to focus on class descriptions, leaving the computation of the class hierarchy to the reasoner, and it can also be used by developers to check if the hierarchy induced by the class descriptions is consistent with their intuition. This may not be the case when, for example, errors in the ontology result in unexpected subsumption inferences, or "under-constrained" class descriptions result in expected inferences not being found. The latter case is extremely common, as it is easy to inadvertently omit axioms that express "obvious" information. For example, an ontology engineer may expect the class of patients who have a fracture of both the tibia and the fibula to be a subClassOf "patient with multiple fractures"; however, this may not be the case if the ontology doesn't include (explicitly or implicitly) the information that the tibia and fibula are different bones. Failure to find the expected subsumption relationship will alert the engineer to the missing DisjointClasses axiom.

Recent work has also shown how reasoning can be used to support modular design [8] and module extraction [7], important techniques for working with large ontologies. When developing a large ontology such as SNOMED, it is useful if not essential to divide the ontology into modules, e.g., to facilitate parallel work by a team of ontology developers. Reasoning techniques can be used to alert the developers to unanticipated and/or undesirable interactions between the various modules. Similarly, it may be desirable to extract from a large ontology a smaller module containing all the information relevant to some subset of the domain, e.g., heart disease—the resulting small(er) ontology will be easier for humans to understand and easier for applications to use. Reasoning can be used to compute a module that is as small as possible while still containing all the necessary information.

Finally, in order to maximise the benefit of reasoning services, tools should be able to explain inferences: without this facility, users may find it difficult to repair errors in the ontology and may even start to doubt the correctness of inferences. Explanation typically involves computing a (hopefully small) subset of the ontology that still entails the inference in question and if necessary presenting the user with a chain of reasoning steps [15]. Figure 5, for example, shows an explanation, produced by the Protégé 4 ontology development tool, of the above mentioned inference with respect to the inconsistency of OceanCrustLayer.

4.2   Reasoning in Deployment

Reasoning is also important when ontologies are deployed in applications—it is needed, e.g., in order to answer structural queries about the domain and to retrieve data. For example, biologists use ontologies such as the gene ontology (GO) and the biological pathways exchange ontology (BioPAX) to annotate data from gene sequencing experiments so as to be able to answer complex queries such as "what DNA binding products interact with insulin receptors". Answering this query requires a reasoner not only to identify individuals that are (perhaps only implicitly) instances of DNA binding products and of insulin receptors, but also to identify which pairs of individuals are (perhaps only implicitly) related via the interactsWith property.

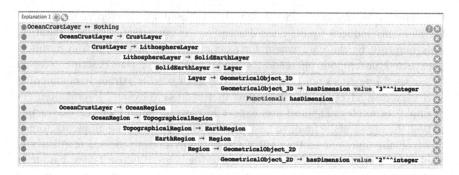

Fig. 5 An explanation from Protégé 4

It is easy to imagine that, with large ontologies, query answering may be a very complex task. The use of DL reasoners allows OWL ontology applications to answer complex queries and to provide guarantees about the correctness of the result. This is obviously of crucial importance when ontologies are used in safety critical applications such as medicine; it is, however, also important if ontology-based systems are to be used as components in larger applications, such as the Semantic Web, where the correct functioning of automated processes may depend on their being able to (correctly) answer such queries.

5 Ontology Applications

The availability of tools and reasoning systems such as those mentioned in Sect. 3 has contributed to the increasingly widespread use of OWL. Applications of OWL are particularly prevalent in the life sciences where it has been used by the developers of several large biomedical ontologies, including the SNOMED, GO and BioPAX ontologies mentioned above, the FMA [5] and the National Cancer Institute thesaurus [10].

Many ontologies are the result of collaborative efforts within a given community and are developed with the aim of facilitating information sharing and exchange. Some ontologies are even commercially developed and subject to a licence fee. In most cases, an ontology focuses on a particular domain, although there are some well-known "foundational" or "upper" ontologies, such as DOLCE[1] and SUMO,[2] whose coverage is more general; their aim is, however, mainly to provide a carefully formalised basis for the development of domain-specific ontologies.

---

[1]http://www.loa-cnr.it/DOLCE.html

[2]http://www.ontologyportal.org/

Many OWL ontologies are now available on the web—an OWL ontology is identified by a URI, and the ontology should, in principle, be available at that location. There are also several well-known ontology libraries, and even ontology search engines such as SWOOGLE,[3] that can be used to locate ontologies. In practice, however, applications are invariably built around a predetermined ontology or set of ontologies that are well understood and known to provide suitable coverage of the relevant domains.

The importance of reasoning support in ontology applications was highlighted by recent work on the MED ontology in which potentially critical errors were discovered (see Sect. 1). Similarly, an extended version of the SNOMED ontology was checked using an OWL reasoner, and a number of missing subClassOf relationships found. This ontology is being used by the UK National Health Service (NHS) to provide "A single and comprehensive system of terms, centrally maintained and updated for use in all NHS organisations and in research" and as a key component of their multi-billion pound "Connecting for Health" IT programme. An important feature of this system is that it can be extended to provide more detailed coverage if needed by specialised applications. For example, a specialist allergy clinic may need to distinguish allergies caused by different kinds of nut, and so may add new terms to the ontology such as AlmondAllergy:

$$\text{AlmondAllergy} \equiv \text{Allergy} \sqcap \exists\text{causedBy.Almond}$$

Using a reasoner to insert this new term into the ontology will ensure that it is recognised as a subClassOf NutAllergy. This is clearly of crucial importance in order to ensure that patients with an AlmondAllergy are correctly identified in the national records system as patients having a NutAllergy.

Ontologies are also widely used to facilitate the sharing and integration of information. The Neurocommons project,[4] for example, aims to provide a platform for sharing and integrating knowledge in the neuroscience domain. A key component is an ontology of annotations that will be used to integrate available knowledge on the web, including major neuroscience databases. Similarly, the OBO Foundry[5] is a library of ontologies designed to facilitate information sharing and integration in the biomedical domain.

In information integration applications the ontology can play several roles: it can provide a formally defined and extensible vocabulary for use in semantic annotations, it can be used to describe the structure of existing sources and the information that they store, and it can provide a detailed model of the domain against which queries can be formulated. Such queries can be answered by using semantic annotations and structural knowledge to retrieve and combine information from

---

[3]http://swoogle.umbc.edu/

[4]http://sciencecommons.org/projects/data/

[5]http://www.obofoundry.org/

multiple sources [26]. It should be noted that the use of ontologies in information integration is far from new and has already been the subject of extensive research within the database community [1].

6 Discussion

Ontology, in its original philosophical sense, is a fundamental branch of meta-physics focusing on the study of existence; its objective is to determine what entities and types of entities actually exist and thus to study the structure of the world. In contrast, in computer science an ontology is an engineering artifact, usually a "conceptual model" of (some aspect of) the world, typically formalised as a logical theory.

Formalising ontologies using logic opens up the possibility of using automated reasoning to aid both their design and deployment. The availability of reasoning tools was an important factor in the design of the OWL ontology language and its basis in description logics and has contributed to the increasingly widespread use of OWL, not only in the Semantic Web per se, but as a popular language for ontology development in diverse application areas. The value of reasoning support has already been demonstrated in medical applications, where it has been used to help repair and enrich ontologies that play an important role in patient care.

The benefits of ontologies and ontology reasoning come at a cost, however. Even with the aid of reasoning enabled tools, developing and maintaining good quality ontologies is a difficult and costly task. Moreover, compared to more established database systems, query answering in an ontology-based information system is likely to be much more difficult and in the worst case could be highly intractable. The use of ontologies is, therefore, perhaps best suited to applications where the schema plays an important role, where it is not reasonable to assume that complete information about the domain is available, and where information has high value.

Although reasoning enabled tools are a boon to ontology engineers, the design of good quality ontologies is still a difficult and time-consuming task, and problems related to the availability of good quality ontologies threaten to limit the deployment of ontology-based information systems. This problem can be tackled to some extent by extending the range and capability of ontology engineering tools, and this is a very active research area. However, tools alone cannot compensate for a lack of understanding of and expertise in "knowledge engineering". This has resulted in ontology engineers increasingly looking to the philosophy community for possible solutions and in particular as a source of relevant expertise in the organisation and formalisation of knowledge. The results of such collaborations are already becoming evident in, for example, the design of foundational ontologies such as DOLCE and SUMO, and in ambitious ontology development projects such as the OBO foundry.

References

1. Batini, C., Lenzerini, M., Navathe, S.B.: A comparative analysis of methodologies for database schema integration. ACM Comput. Surv. **18**(4), 323–364 (1986)
2. Brachman, R.J., Schmolze, J.G.: An overview of the KL-ONE knowledge representation system. Cognit. Sci. **9**(2), 171–216 (1985)
3. Derriere, S., Richard, A., Preite-Martinez, A.: An ontology of astronomical object types for the virtual observatory. In: Proceedings of Special Session 3 of the 26th Meeting of the IAU: Virtual Observatory in Action: New Science, New Technology, and Next Generation Facilities (2006)
4. Fensel, D., van Harmelen, F., Horrocks, I., McGuinness, D., Patel-Schneider, P.F.: OIL: an ontology infrastructure for the semantic web. IEEE Intell. Syst. **16**(2), 38–45 (2001)
5. Golbreich, C., Zhang, S., Bodenreider, O.: The foundational model of anatomy in OWL: experience and perspectives. J. Web Semant. **4**(3), 181–195 (2006)
6. Goodwin, J.: Experiences of using OWL at the ordnance survey. In: Proceedings of the First OWL Experiences and Directions Workshop. CEUR Workshop Proceedings, vol. 188, CEUR Online Workshop Proceedings (http://ceur-ws.org/) (2005)
7. Grau, B.C., Horrocks, I., Kazakov, Y., Sattler, U.: Just the right amount: extracting modules from ontologies. In: Proceedings of the Sixteenth International World Wide Web Conference (WWW 2007). ACM Press, New York (2007a)
8. Grau, B.C., Kazakov, Y., Horrocks, I., Sattler, U.: A logical framework for modular integration of ontologies. In: Proceedings of the 20th International Joint Conference on Artificial Intelligence (IJCAI 2007), pp. 298–303 (2007b)
9. Haarslev, V., Möller, R.: RACER system description. In: Proceedings of the International Joint Conference on Automated Reasoning (IJCAR 2001). Lecture Notes in Artificial Intelligence, vol. 2083, pp. 701–705. Springer, Berlin (2001)
10. Hartel, F.W., de Coronado, S., Dionne, R., Fragoso, G., Golbeck, J.: Modeling a description logic vocabulary for cancer research. J. Biomed. Informat. **38**(2), 114–129 (2005)
11. Horrocks, I., Sattler, U.: A tableaux decision procedure for $\mathcal{SHOIQ}$. In: Proceedings of the 19th International Joint Conference on Artificial Intelligence (IJCAI 2005), pp. 448–453. Professional Book Center. AAAI Press, Menlo Park (2005)
12. Horrocks, I., Patel-Schneider, P.F., van Harmelen, F.: Reviewing the design of DAML + OIL: an ontology language for the semantic web. In: Proceedings of the 18th National Conference on Artificial Intelligence (AAAI 2002), pp. 792–797. AAAI Press (2002), ISBN 0-26251-129-0
13. Horrocks, I., Patel-Schneider, P.F., van Harmelen, F.: From $\mathcal{SHIQ}$ and RDF to OWL: the making of a web ontology language. J. Web Semant. **1**(1), 7–26 (2003) ISSN 1570–8268
14. Kalyanpur, A., Parsia, B., Sirin, E., Cuenca-Grau, B., Hendler, J.: SWOOP: a web ontology editing browser. J. Web Semant. **4**(2) (2005a)
15. Kalyanpur, A., Parsia, B., Sirin, E., Hendler, J.: Debugging unsatisfiable classes in OWL ontologies. J. Web Semant. **3**(4), 243–366 (2005b)
16. Kershenbaum, A., Fokoue, A., Patel, C., Welty, C., Schonberg, E., Cimino, J., Ma, L., Srinivas, K., Schloss, R., Murdock, J.W.: A view of OWL from the field: use cases and experiences. In: Proceedings of the Second OWL Experiences and Directions Workshop. CEUR Workshop Proceedings, vol. 216 (http://ceur-ws.org/). CEUR (2006)
17. Knublauch, H., Fergerson, R., Noy, N., Musen, M.: The Protégé OWL Plugin: an open development environment for semantic web applications. In: McIlraith, S.A., Plexousakis, D., van Harmelen, F. (eds.) Proceedings of the 3rd International Semantic Web Conference (ISWC 2004). Lecture Notes in Computer Science, vol. 3298, pp. 229–243. Springer, Berlin (2004), ISBN 3-540-23798-4
18. Lacy, L., Aviles, G., Fraser, K., Gerber, W., Mulvehill, A., Gaskill, R.: Experiences using OWL in military applications. In: Proceedings of the First OWL Experiences and Directions Workshop. CEUR Workshop Proceedings, vol. 188 (http://ceur-ws.org/). CEUR (2005)

19. Patel-Schneider, P.F., Hayes, P., Horrocks, I.: OWL Web Ontology Language semantics and abstract syntax, W3C Recommendation, 10 February 2004, URL http://www.w3.org/TR/owl-semantics/
20. Rector, A., Rogers, J.: Ontological and practical issues in using a description logic to represent medical concept systems: experience from GALEN. In: Reasoning Web, Second International Summer School, Tutorial Lectures. LNCS, vol. 4126, pp. 197–231. Springer, Heidelberg (2006)
21. Ruttenberg, A., Rees, J., Luciano, J.: Experience using OWL DL for the exchange of biological pathway information. In: Proceedings of the First OWL Experiences and Directions Workshop. CEUR Workshop Proceedings, vol. 188 (http://ceur-ws.org/). CEUR (2005)
22. Sidhu, A., Dillon, T., Chang, E., Sidhu, B.S.: Protein ontology development using OWL. In: Proceedings of the First OWL Experiences and Directions Workshop. CEUR Workshop Proceedings, vol. 188 (http://ceur-ws.org/). CEUR (2005)
23. Sirin, E., Parsia, B., Cuenca Grau, B., Kalyanpur, A., Katz, Y.: Pellet: a practical OWL-DL reasoner. J. Web Semant. 5(2), 51–53 (2007)
24. Soergel, D., Lauser, B., Liang, A., Fisseha, F., Keizer, J., Katz, S.: Reengineering thesauri for new applications: the AGROVOC example. J. Digital Inform. 4(4) (2004)
25. Spackman, K.: Managing clinical terminology hierarchies using algorithmic calculation of subsumption: experience with SNOMED-RT. J. Am. Med. Informat. Assoc. (2000) Fall Symposium Special Issue
26. Stevens, R., Baker, P., Bechhofer, S., Ng, G., Jacoby, A., Paton, N.W., Goble, C.A., Brass, A.: Tambis: transparent access to multiple bioinformatics information sources. Bioinformatics 16(2), 184–186 (2000)
27. SWEET: Semantic web for earth and environmental terminology (SWEET). Jet Propulsion Laboratory, California Institute of Technology (2006), http://sweet.jpl.nasa.gov/
28. Tsarkov, D., Horrocks, I.: FaCT++ description logic reasoner: system description. In: Proceedings of the International Joint Conference on Automated Reasoning (IJCAR 2006). Lecture Notes in Artificial Intelligence, vol. 4130, pp. 292–297. Springer, Berlin (2006)
29. Woods, W.A.: What's in a link: foundations for semantic networks. In: Brachman, R.J., Levesque, H.J. (eds.) Readings in Knowledge Representation, pp. 217–241. Morgan Kaufmann Publishers, San Francisco, California (1985); Previously published In: Bobrow, D.G., Collins, A.M. (eds.) Representation and Understanding: Studies in Cognitive Science, pp. 35–82. Academic, New York (1975)

Taxonomic Change as a Reflection of Progress in a Scientific Discipline

Alexa T. McCray and Kyungjoon Lee

Abstract The terminology that is used to index the world's biomedical literature has its roots in an indexing system developed in the late nineteenth century. In the early 1960s that terminology, now known as the Medical Subject Headings (MeSH), was updated and reframed for information retrieval purposes in the context of MEDLINE, the US National Library of Medicine's bibliographic retrieval system. MeSH is a semantic system in its own right and responds, just as all such systems do, to both internal and external forces. We conducted a study of the evolution of MeSH over the last 45 years. We hypothesized, and our analyses confirmed, that some changes reflect internal considerations, such as developing a more principled ontological structure, and others reflect important external forces, most notably the development of biomedical knowledge itself. Our work has implications for research in ontology evolution and for research that is concerned with the conceptual modeling and evolution of a knowledge domain.

1 Introduction

The terminology that is used to index the world's biomedical literature has its roots in an indexing system developed in the late nineteenth century [1]. In the early 1960s that terminology, known as the Medical Subject Headings (MeSH), was updated and reframed for information retrieval purposes in the context of MEDLINE, the US National Library of Medicine's bibliographic retrieval system. MeSH began as a loose collection of indexing terms and has over the years grown to become a highly structured arrangement of tens of thousands of terms that are used to index hundreds of thousands of articles per year. MeSH is technically a thesaurus [2], though it is also increasingly referred to as a biomedical ontology [3–5]. For the purposes of the

A.T. McCray (✉) · K. Lee
Center for Biomedical Informatics, Harvard Medical School, Boston, MA, USA
e-mail: alexa_mccray; joon_lee@hms.harvard.edu

B.-O. Küppers et al. (eds.), *Evolution of Semantic Systems*,
DOI 10.1007/978-3-642-34997-3_10, © Springer-Verlag Berlin Heidelberg 2013

following discussion we will consider MeSH to be an ontology in Gruber's sense, that is, as a conceptualization of the biomedical domain [6, 7].

As a continually evolving ontology, MeSH responds, as all such systems must, to both internal and external forces. For this initial study, we conducted an analysis of the psychiatric and psychologic terms, the so-called "F" section in MeSH, and studied the evolution of this section over the last 45 years, in 5-year increments. The purpose of the study reported here is to assess the degree to which changes in MeSH reflect the evolution of biomedical knowledge as this is revealed in the biomedical literature. We hypothesize that both structural and terminological changes in MeSH are involved. Our work is being conducted as part of a larger study in identifying and analyzing research trends in the biomedical domain. To that end, we have implemented Medvane, an automated, interactive system for visualizing developments in biomedical research.

In the remainder of the introduction we provide brief background information about MeSH, MEDLINE, and Medvane. We conclude the introduction with a review of key concepts in the evolution of ontologies. Next, we present our methods and materials, our results, and we end with a discussion of several examples of significant changes in MeSH that clearly reflect the evolution of biomedical knowledge.

1.1   Medical Subject Headings

MeSH is the controlled terminology designed in the 1960s by the National Library of Medicine (NLM) for the purpose of indexing and retrieving the biomedical literature. The scope of MeSH is very broad (all of biomedicine) and, according to Nelson et al. the goal of the terminology is "to provide a reproducible partition of concepts relevant to biomedicine for purposes of organization of medical knowledge and information." [8].

MeSH currently contains over 26,000 descriptors, arranged in a hierarchical structure of 16 top-level categories with up to 11 levels of embedding. MeSH top-level categories include specific topics such as anatomy, organisms, diseases, drugs, and procedures, as well as categories that cover large disciplinary areas, such as anthropology, education, and sociology, technology, industry, and agriculture, humanities, and information science. There are also some 180,000 "entry terms," which are synonyms or near-synonyms. For example, "Antisocial Personality Disorder" has five entry terms, one is a lexical variant, "Personality Disorder, Antisocial," and the others are closely related terms, "Antisocial Personality," "Dyssocial Behavior," "Psychopathic Personality," and "Sociopathic Personality." The entry terms share the same unique identifier (DUID) with the MeSH descriptor so that searching with an entry term gives the same search results as searching with the preferred descriptor name.

Each MeSH descriptor also has one or more tree numbers, representing its place in the MeSH hierarchy. Any given descriptor can appear in multiple hierarchies, according to a particular perspective on its meaning. For example, "Lung Neoplasms" appears both in the "Neoplasms" hierarchy (C04) and in the "Respiratory Tract Diseases" (C08) hierarchy.

MeSH is updated annually with changes both in its terminology and in its hierarchical structure. Sometimes the changes to MeSH are quite minor, e.g., a singular form becomes a plural, or a generic term is substituted for an eponym, and in other cases the changes are more significant. In modifying MeSH, NLM subject matter specialists are primarily guided by the literature itself, but they also track developments in a wide range of specialized domain terminologies. MeSH staff characterize changes to MeSH as follows:

"Before new descriptors are introduced, there is careful consideration of how the concept is currently indexed or cataloged. If the existing descriptors ... precisely characterize or identify the literature on the subject, there may not be a need for a new descriptor. Both too much change or too little change are to be avoided as MeSH is kept current with changes in biomedical knowledge." [9]

1.2 MEDLINE

MEDLINE is an extensive database of citations to journal articles in the life sciences produced by the NLM. MEDLINE currently comprises over 20 million records covering the broad field of biomedicine and health, including medicine, nursing, dentistry, veterinary medicine, health care systems, and preclinical sciences. Over 5,500 scholarly journals from around the world are represented. Citations are added every day and contain standard bibliographic metadata, such as title, author, and source as well as publication type (e.g., review, case report, meta-analysis), grant numbers, and, importantly, MeSH terms. In a large percentage of cases, author-created abstracts are also available. See Fig. 1 for a sample MEDLINE citation record.

Each citation record in MEDLINE is indexed with a set of MeSH terms that characterizes the full article. Although citation records include only abstracts, trained indexers review the full text of articles during the indexing process. Typically, between three to five MeSH terms describe the major topics discussed in the article. Additional MeSH terms indicate other topics that are relevant to the article, but which are not of primary importance for capturing the core content.

MEDLINE can be searched through the PubMed search interface using MeSH terms, author names, title words, text words or phrases, journal names, or any combination of these. Additional, more complex searches may be done by specifying date fields, age groups, gender, or human or animal studies.

While citation records are added daily, at the end of each calendar year, so-called file maintenance is performed on the MEDLINE database [10, 11]. A new database is created that entirely replaces the previous year's version. In practice, what this means is that when a new MeSH term replaces an older one in the MeSH vocabulary, then the new term is substituted for the older one in each citation record. In most

Science. 2004 Apr 16;304(5669):438-41. Epub 2004 Mar 18.

Integration of word meaning and world knowledge in language comprehension.

Hagoort P, Hald L, Bastiaansen M, Petersson KM.

Abstract
Although the sentences that we hear or read have meaning, this does not necessarily mean
that they are also true. Relatively little is known about the critical brain structures for, and
the relative time course of, establishing the meaning and truth of linguistic expressions. We
present electroencephalogram data that show the rapid parallel integration of both semantic
and world knowledge during …

MeSH Terms
Adult | Brain Mapping | Comprehension* | Electroencephalography | Evoked Potentials |
Female | Humans | Knowledge* | Language* | Linguistics* | Magnetic Resonance Imaging |
Male | Memory/physiology | Prefrontal Cortex/physiology* | Semantics*

PMID: 15031438 [PubMed - indexed for MEDLINE]

Fig. 1 Sample MEDLINE citation record. MeSH terms with an asterisk are major topics of the
article

cases, when this happens, the older term becomes an entry term to the newer one in
MeSH, thus, ensuring that if users search under the older term, they will still retrieve
the relevant articles.

Both the annual MeSH and MEDLINE files are freely available for download in
XML format [12].

1.3   Literature Trends

There is growing interest in detecting and visualizing trends in the scientific
literature [13–21]. Our own work concerns the identification and analysis of
research trends in the broad biomedical domain. We have implemented Medvane, an
automated, interactive system for visualizing developments in biomedical research,
as these are represented in MEDLINE.

Any given article appears in a specific journal, has one or more authors, is funded
by a specific funding agency, is indexed by a set of MeSH terms, etc. Medvane
allows users to focus on a specific journal and note the most frequent topics, authors,
and publication types that have been associated with that journal over time. Or, it is
possible to view, according to specific time periods, an author's publication history,
including the top journals in which that author has published, the most significant
topics studied by that author, and the author's primary co-authors. Medvane presents
the relationships between all of these aspects and their evolution over time. We
initially undertook the research reported in this paper in order to, among other
things, assess the impact of the evolution of the MeSH ontological structure on the
interpretation of historical trends in biomedical research.

1.4   Ontology Evolution

In the context of recent research on the Semantic Web, there has been renewed interest in ontology evolution, with a focus on versioning and change management [22–26]. It has become clear that ontologies are not static artifacts, but rather, just like the knowledge they encapsulate, ontologies necessarily evolve. The approach to ontology evolution in these cases is to develop methods for recognizing both terminological change and structural change with the explicit goal of ensuring continued (inter)operability. In some cases, tools have been developed for revealing the differences between various versions of an ontology such that ontology developers are able to manage modifications as the ontology grows.

While insights from ontology versioning methods are pertinent to the work we describe here, our specific techniques for assessing the evolution of MeSH are closely related to research on ontology alignment and mapping [27–29].[1] Euzenat and Shvaiko propose several levels (and associated techniques) at which the ontology mapping process may occur [27]. Following the work of Kang and Naughton [30], they make a high level distinction between element level analysis, on the one hand, encompassing a variety of string, name-based, and linguistic techniques for assessing changes in terminology, and structure-level analysis, on the other hand, including taxonomic and topological similarity measures for assessing changes in overall structure.

2   Methods and Materials

2.1   Materials

We created a dataset comprising the F (psychiatry and psychology) tree sections of MeSH in 5-year increments over a 45 year period, from 1963 to 2008. Because early versions of MeSH are only available in print form, we reconstructed the data from the printed tree structure books for the time periods 1963–1988. Each digitally reconstructed file included the name of the term and its place in the hierarchy. For the more recent time period, we received the data in electronic form from the National Library of Medicine.[2] In all cases, the only data we had access to were the terms themselves and the implicit or explicit tree structures. Full MeSH records as they exist today are not, in general, available for previous years. In order to take advantage of more extensive data about MeSH descriptors, we downloaded the 2008 MeSH from NLM. Table 1 shows some of the information that is available in a 2008 MeSH record. The 2008 DUID and 2008 entry terms were used in all of our analyses.

---

[1] This is not remarkable if we view each of our ten slices of MeSH as separate (though, of course, closely related) ontologies which are then aligned with each other across the years.

[2] Notably, in 2009 the printed MeSH was discontinued.

Table 1 Partial 2008 MeSH record

MeSH Heading	Dementia
DUID	D003704
Tree numbers	C10.228.140.380; F03.087.400
Scope note	An acquired organic mental disorder'
Entry terms	Amentia; Familial Dementia; Frontotemporal Lobar Degeneration; Semantic Dementia; Senile Paranoid Dementia

2.2 Methods

Building Hierarchy from Implicit Structures While the presentation is primarily alphabetical in the early printed MeSH books, there is an additional implied hierarchy. Consider Fig. 2, a printed page from MeSH 1973.

Terms on the left margin are all in uppercase. Indented under some of these are additional terms in mixed case. Consider, for example, the MeSH term *behavior* in Fig. 2. Indented under it is the term "Behavior, Animal." Looking further down the page *behavior, animal* is represented again in all uppercase with further terms, e.g., "Animal Communication," indented under it. *animal communication* also appears listed in uppercase in the left-hand column, and it likewise has several terms indented under it. Thus, there is a four-level implied hierarchy here, e.g., "Vocalization, Animal" is a child of *animal communication*, which is a child of (indented under) *behavior, animal*, which is itself a child of *behavior*. Further, some terms have multiple implied parents, e.g., "Animal Communication" is a child of both *behavior, animal* and *communication*.

Occasionally, the implied structure led to an infinite loop in our recursive tree building process. This happened when a term was both a child and a parent of another term. This was rare and only occurred in the very early years. In these cases we stopped the recursion after two levels of embedding.

By 1978 the modern tree number system had been introduced, and in the printed versions MeSH terms are displayed hierarchically together with their (sometimes multiple) tree numbers. See Fig. 3.

Because we were interested not only in the evolution of individual MeSH terms but also in the changes of the MeSH structure itself, we assigned (arbitrary) tree numbers to the terms in the earlier years (1963, 1968, and 1973), and, for comparability, we normalized the original MeSH tree numbers (assigning similar arbitrary tree numbers) for the later years (1978–2008). We assigned the F category root node the number "6." (It is the sixth top-level category in MeSH.) We then assigned consecutive numbers starting from "1" in the order the child nodes appear and repeated the same process for all of their descendant nodes. For example, in 2008 "Behavior and Behavior Mechanisms" has the MeSH tree number "F01." This is normalized as "6.1" (each dot indicates one level of embedding) in our scheme. "Attitude," which has the MeSH tree number "F01.100," is the second child of

ABSENTEEISM (J)	BEHAVIOR	BOREDOM
ACHIEVEMENT	Aggression (F2)	CEREMONIAL (I)
ACTING OUT (F2, F3)	Attitude	CHARACTER
ADAPTATION, PSYCHOLOGICAL	Autism (F2)	CHILD BEHAVIOR
AFFECT	Behavior, Animal	CHILD DEVELOPMENT
AFTERIMAGE	Ceremonial (I)	Language Development
AGGRESSION (F2)	Child Behavior	CHILD DEVELOPMENT
ALCOHOL DRINKING	Communication (L)	DEVIATIONS (F2)
ANGER	Competitive Behavior	Enuresis (C6, F2)
	Compulsive Behavior (F2)	Fingersucking (F2)
ANIMAL COMMUNICATION	Cooperative Behavior	Nail Biting (F2)
Pheromones (D12)	Displacement (Psychology) (F2)	Tongue Habits (F2)
Vocalization, Animal	Drinking Behavior	CHILD, GIFTED (M)
ANOMIE (I)	Exploratory Behavior	CHILD GUIDANCE (F3)
Social Alienation (F2)	Feeding Behavior	CHILD REARING
ANXIETY	Gambling (F2)	Toilet Training
Anxiety, Separation (F2)	Habits	
ANXIETY, SEPARATION (F2)	Imitative Behavior	COGNITION
APPETITE (G1)	Impulsive Behavior (F2)	Cognitive Dissonance
APPETITIVE BEHAVIOR	Inhibition (Psychology) (F2)	Consciousness
APTITUDE	Maternal Behavior	Imagination
	Motor Activity (G1)	Perception
AROUSAL (G1)	Paternal Behavior	Thinking
Attention	Self Stimulation	COGNITIVE DISSONANCE
Wakefulness (G1)	Sex Behavior	COITUS (G1)
	Social Behavior	COLOR PERCEPTION (G1)
ASSOCIATION (F3)	Social Conformity	
ATTENTION	Social Dominance	COMMUNICATION (L)
ATTITUDE	Social Facilitation	Animal Communication
Attitude of Health Personnel (N4)	Spatial Behavior (I)	Crying (C17)
Attitude to Health (N1)	Sucking Behavior	Facial Expression (E1)
Personal Satisfaction		Kinesics
Set (Psychology)	BEHAVIOR, ANIMAL	Language (L)
ATTITUDE OF HEALTH	Animal Communication	Nonverbal Communication
PERSONNEL (N4)	Appetitive Behavior	Persuasive Communication
ATTITUDE TO HEALTH (N1)	Consummatory Behavior	Speech (G1, L)

Fig. 2 Section of a printed page from MeSH 1973

BEHAVIOR AND BEHAVIOR MECHANISMS (NON MESH)	F1	
ADAPTATION, PSYCHOLOGICAL	F1.58	
ORIENTATION	F1.58.577	F2.830.606
BEHAVIOR	F1.145	
ACCIDENT PRONENESS	F1.145.15	
AGGRESSION	F1.145.32	F3.126.842.
AGONISTIC BEHAVIOR ·	F1.145.32.268	
ATTITUDE	F1.145.76	F1.829.92
ATTITUDE OF HEALTH PERSONNEL	F1.145.76.147	F1.829.92.
ATTITUDE TO DEATH	F1.145.76.247	
ATTITUDE TO HEALTH	F1.145.76.347	F1.829.92.
PATIENT ACCEPTANCE OF HEALTH CARE	F1.145.76.347.570	
PERMISSIVENESS ·	F1.145.76.675	

Fig. 3 Section of a printed page from MeSH 1978

"Behavior and Behavior Mechanisms" and is, thus, assigned the normalized tree number "6.1.2." This tree number assignment and normalization preserves the order within the same year and makes the comparison of the node position of a given term across different years possible.

Term-Level Changes So that we would be able to study and explore the structural and terminological changes in MeSH over time, we built an interactive web application, which we call the MEVO (MeSH Evolution) Browser. MEVO has three

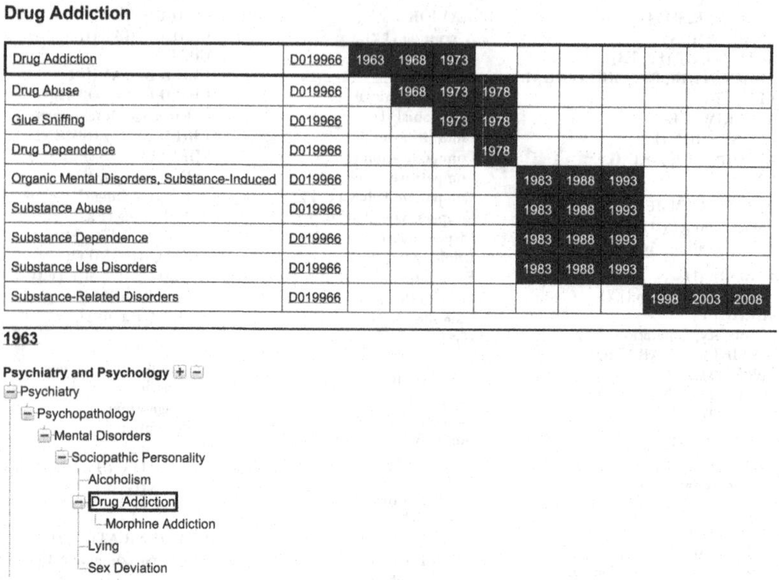

Drug Addiction

Drug Addiction	D019966	1963	1968	1973							
Drug Abuse	D019966		1968	1973	1978						
Glue Sniffing	D019966			1973	1978						
Drug Dependence	D019966				1978						
Organic Mental Disorders, Substance-Induced	D019966					1983	1988	1993			
Substance Abuse	D019966					1983	1988	1993			
Substance Dependence	D019966					1983	1988	1993			
Substance Use Disorders	D019966					1983	1988	1993			
Substance-Related Disorders	D019966								1998	2003	2008

1963

Psychiatry and Psychology ⊞ ⊟
⊟ Psychiatry
  ⊟ Psychopathology
    ⊟ Mental Disorders
      ⊟ Sociopathic Personality
        | Alcoholism
        ⊟ Drug Addiction
          └ Morphine Addiction
        | Lying
        └ Sex Deviation

Fig. 4 MEVO screenshot: the evolution of DUID D19966 from 1963–2008

distinct views: a tree structure view, a year-by-year view, and a term view, grouped by DUID. These three views are fully hyperlinked with each other so that traversing from one view to another can be done easily. Figure 4 shows a MEVO screenshot.

The screenshot shows the evolution of the descriptor, DUID D019966, from 1963 to 2008. In 1963 only "Drug Addiction" appeared in MeSH. In 1968 "Drug Abuse" was added, and in 1973 "Glue Sniffing" joined this group of co-existing MeSH descriptors. By 1978, "Drug Addiction" had been dropped, and "Drug Dependence" was introduced. From 1983 to 1993 a different set of four descriptors co-existed and all of the earlier ones were dropped. In 2008 the official name of DUID D019966 is "Substance-Related Disorders" and each of the other eight historical terms have become entry terms. For any given year the terms listed in Fig. 4 had their own DUID's. For example, "Organic Mental Disorders, Substance-Induced," "Substance Abuse," "Substance Dependence," and "Substance Use Disorders" co-existed in 1993 with four different DUID's. By 1998 they were all merged with "Substance-Related Disorders" and they now share one DUID. DUID's are never reused, but they may be deleted.

Just below the term display in Fig. 4, a portion of the fully navigable tree structure is shown, with a focus here on the term "Drug Addiction." Scrolling down the display reveals the tree structures for each of the years when "Drug Addiction" existed as an MeSH descriptor (1963–1973). MEVO is fully interactive. Clicking on any term gives a view that focuses specifically on that term and its historical evolution. Clicking on a year displays the tree structures, terms, and additions and deletions for that year and is an entry point for any of the other years.

Global Changes

Terminology When comparing MeSH terms across multiple years we did so at the string, lexical, and descriptor levels. That is, our comparisons looked, respectively, for the absence or presence of

1. The same string of characters
2. The same term, but where the actual strings might be different because of lexical variation[3]
3. Different terms, but where the terms have the same MeSH descriptor DUID[4]

The results at each of these different levels can be quite different. For example, in 1978 an MeSH descriptor was expressed as "Psychiatry, Community," and by 1983 it had become "Community Psychiatry." So, while the strings are different, they only vary by word order. We would not consider this either as a deletion or as an addition at the lexical level. At the string level, however, we would say the 1978 string had been deleted and the 1983 string had been added. Similarly, in 1973 a MeSH descriptor was expressed as "Mongolism", and by 1978 it had become "Down's Syndrome". These are completely different strings and terms and thus at the string and lexical levels they would be considered deletions and additions, respectively. However, at the descriptor level, because they share the same DUID they are counted as equivalent.

Structure From one year to another across our ten time slices, we compared descriptor additions and deletions, changes in depth and breadth of hierarchies, and movement of descriptors within hierarchies. We distinguished true deletions, i.e., when a string is entirely deleted from MeSH, from those cases when a string is no longer a descriptor in its own right, but rather has become an entry term to a different descriptor. We introduced the notion of "demotion" to account for these latter cases. To measure structural change, we used our normalized tree numbers and adopted Resnik's revision [34] of Wu–Palmer's similarity measure [35]. Similarity s between terms c and c' is defined as follows:

$$s(c, c') = 2 \times p(c \wedge c')/(p(c) + p(c')), \qquad (1)$$

where $p(c)$ is the number of edges from the root node to c and $c \wedge c'$ is the most common ancestor of c and c'. Similarity s becomes 1 when c and c' are the same (i.e., when there has been no movement) and approaches 1 when there is a minor movement, for example, from one deeply embedded node to another where the

---

[3]We used the UMLS lexical programs, and specifically the "norm" program to evaluate lexical variation [31,32].

[4]These distinctions are similar, but not identical, to the sui (string), lui (lexical variant), and cui (concept) distinctions in the Unified Medical Language System [33]. The DUID often functions like a concept, where its entry terms are synonymous with each other, but just as often the DUID includes a combination of synonymous entry terms and terms that are only closely related to each other.

nodes share a proximal ancestor. Similarity approaches zero when both c and c' are deeply embedded and their common ancestor is close to the root. For example, if a term were to move from being a deeply embedded descendant of "Behavior and Behavior Mechanisms" (which lies close to the root) in 1998 to become a descendant of Mental Disorders (which also lies close to the root) in 2003, this would signal a significant change in interpretation of that term. Because distance and similarity are inversely related to each other (the greater the distance the less similar the terms are), we defined distance, d between terms c and c' as follows:

$$d(c, c') = 1 - s(c, c'). \qquad (2)$$

Distance d becomes zero when c and c' are the same, and approaches 1 when c and c' are further apart.

An example illustrates. "Peer Review" has the same MeSH tree number in 1998 and 2003, but two different normalized tree numbers, "6.1.12.2.4" and "6.1.13.3.5," respectively. This means that new terms were added in 2003 at the grandparent and parent levels as well as at the same level. (Recall that each dot represents one level of embedding.) The distance between the position of "Peer Review" in 1998 and its new position in 2003 is calculated as follows:

$$d = 1 - 2 \times p(6.1)/(p(6.1.12.2.4) + p(6.1.13.3.5)) \qquad (3)$$
$$= 1 - 2 \times 1/(4 + 4)$$
$$= 1 - 0.25$$
$$= 0.75.$$

When a term is added or deleted from one time period to the next, we define distance, d as follows:

$$d(c) = 1/p(c) \times w, \qquad (4)$$

where $p(c)$ is the number of edges from the root node to c and w is the weight factor.

We assigned a weight factor of 3 for both addition and deletion, and a weight factor of 2 for demotion. We gave greater weight to additions and deletions because they have a greater impact on the overall vocabulary. For example, "Substance-Related Disorders" was added in 1998 as F03.900 (6.3.16), where the distance is $\frac{1}{2} \times 3 = 1.5$. Demotions, on the other hand, reflect an internal change where a descriptor is not fully removed, but, rather, is incorporated within another existing descriptor as an entry term. For example, "Dementia, Presenile" and "Alzheimer's Disease" co-existed as descriptors until 1993, and then became entry terms of "Alzheimer's Disease" in 1998. The distance for "Dementia, Presenile" (normalized tree number 6.3.5.4.1.4.2) between 1993 and 1998 is calculated as follows (with the demotion weight of 2):

$$d = 1/p(6.3.5.4.1.4.2) \times 2 = 0.33. \qquad (5)$$

Finally, when there is more than one tree number for a given term in the same year, we calculated all possible pairs of those tree numbers across years and (arbitrarily) chose the minimum distance. The overall structural change between two years is the sum of all distances for each individual term. To compare this change with other year pairs, we divided the sum of all distances for each term across the two years by the total number of unique terms that existed in these two years.

Literature We used the Entrez programming utilities [36] to retrieve all of the articles indexed with descriptors from the F section of MeSH in each of our 10 time periods and also the full set of articles published during each of those same periods. So as not to overestimate the relative growth of the psychiatry and psychology literature over time, we normalized the growth of articles by dividing the number of articles indexed by descriptors from the F section of MeSH by the total number of articles published during any given time period.

3 Results

3.1 Term-Level Changes

Using the MEVO browser, we explored changes in individual MeSH terms or groups of MeSH terms over time. Tables 2 and 3, for example, show the evolution of term-level changes for two interesting cases, autism spectrum disorders (ASD) and human sexuality.

3.2 Global Changes

Changes in Terminology The total number of descriptors in the "F" section of MeSH in 1963 was 268. By 2008 the total number had increased to 858. Figure 5 represents the changes in the number of terms, at the lexical level, since 1963, represented as additions and deletions.

Figure 5 shows that there was significant growth in the early years, until about 1978. Thereafter, the growth has been moderate, with a dip in 1998. In total, over the last 45 years, 974 descriptors were added to the F section of MeSH and 384 terms were deleted as descriptors. In most cases these deleted descriptors were retained as entry terms.

Changes in Structure Table 4 shows the results of the calculations described above and represents the overall structural change in the F section of MeSH since 1963.

The first row gives the sum of the structural distances for each individual MeSH term in the F section from one time period to another. The second row gives the total number of unique terms that existed in the time period, and the third row gives the mean of the structural distances.

Table 2 Changes in autism terminology over the years

	'63	'68	'73	'78	'83	'88	'93	'98	'03	'08	08 DUI
Autism	●	●	●	●	●	●	●	●			D001321
Autism, early infantile			●	●							D001321
Autism, infantile					●	●	●	●			D001321
Autistic disorder									●	●	D001321
Rett syndrome						●				●	D015528
Asperger syndrome									●	●	D020817
Childhood schizophrenia	●	●	●	●	●	●	●	●	●	●	D012561
PCDD					●	●	●	●	●	●	D002659

Table 3 Changes in sexuality terminology over the years

	'63	'68	'73	'78	'83	'88	'93	'98	'03	'08	08 DUI
Homosexuality	●	●	●	●	●	●	●	●	●	●	D006716
Homosexuality, female				●				●	●	●	D018452
Homosexuality, ego-dystonic					●	●	●				D006716
Bisexuality						●	●	●	●	●	D001727
Sexuality								●	●	●	D019529
Heterosexuality								●	●	●	D020010
Homosexuality, male								●	●	●	D018451

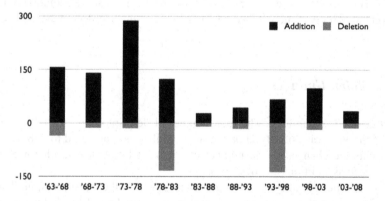

Fig. 5 Total number of additions and deletions of psychiatry and psychology terms by year

Table 4 Structural change in the F section of MeSH

	'63–'68	'68–'73	'73–'78	'78–'83	'83–'88	'88–'93	'93–'98	'98–'03	'03–'08
(a)	339	462	624	371	53	150	399	282	92
(b)	424	529	803	913	803	837	892	854	876
(c)	0.799	0.873	0.777	0.406	0.066	0.179	0.447	0.330	0.105

(a) Sum of structural distances; (b) Number of terms; (c) Mean of structural distances

Table 5 Growth in MEDLINE citations over the years

	'63–'68	'68–'73	'73–'78	'78–'83	'83–'88	'88–'93	'93–'98	'98–'03	'03–'08
(a)	12,484	6,942	4,518	5,102	9,774	9,950	12,199	26,897	34,551
(b)	64,287	23,048	39,778	35,269	75,314	37,661	49,437	120,612	217,996
(c)	0.194	0.301	0.113	0.145	0.130	0.264	0.247	0.223	0.158

(a) Increased number of articles in F section; (b) increased number of articles in MEDLINE; (c) normalized increased number of articles in F section (a / b)

Growth in Literature Table 5 shows the growth in the literature over the years.

The table shows the increase in the number of MEDLINE citations indexed by MeSH F section terms normalized by the overall increase in citations over the years. Row c in the table indicates a burst of research in psychiatry and psychology between 1968 and 1973 and then again between 1988 and 1993.

4 Discussion

Our study of the psychiatric and psychologic terms in MeSH revealed a large number of both structural and terminological changes in the last 45 years. We found that some changes reflect internal considerations, such as developing a more principled ontological structure. This was particularly evident in the early years when the vocabulary was beginning to take shape. The 1963 version had very little explicit structure; by 1973 three top-level categories had been introduced; and by 1978 a fully formed hierarchical structure with tree numbers had been established. In the early years there was one top-level F node, "psychiatry and psychology." Table 6 shows that by 1973, there were three top-level nodes, by 1978 there were four, and in the last decade there have been four slightly different top-level nodes.

The changes in the top-level categories between 1993 and 1998 represent a clear ontological "clean-up," the result of which was that behaviors were clearly separated from disorders. In addition, between 1993 and 1998 a large number of terms were deleted from the F section. (See Fig. 5 above.) In many cases, as has already been noted, the deleted term did not disappear from MeSH entirely, but rather, it survived as an entry term. In some other cases, the deleted term was moved to a different section of MeSH, e.g., "decision theory," which had been a child of "psychological techniques," was moved outside of F to two different MeSH categories (H and L, where it appears under Mathematical Computing) This drop in the number of terms in 1998 is another indicator of a thorough review and revision of the psychiatry and psychology section in this time period.

The impact of these changes, and others at more granular levels, is not to be underestimated. Search algorithms depend on the structure of MeSH when performing queries, and an ill-formed structure will lead to incomplete results at best. In the past, (expert) users were more aware of the structure of MeSH when formulating queries, and could, perhaps, better control the search results. Today, the

Table 6 Top-level nodes over time in the psychiatry and psychology section of MeSH

Time periods	Top-level categories
1973	Behavioral sciences, psychological tests, psychotherapy, services
	Behavioral symptoms and mental disorders
	Psychologic mechanisms and processes
1978–1993	Behavior and behavior mechanisms
	Behavior and mental disorders
	Disciplines, tests, therapy, services
	Psychologic processes and principles
1998–2008	Behavior and behavior mechanisms
	Behavioral disciplines and activities
	Mental disorders, psychological phenomena and processes

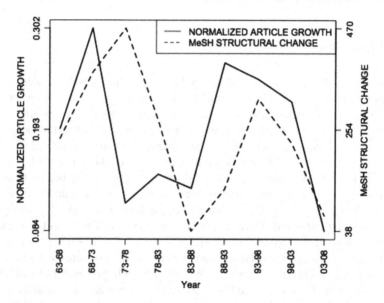

Fig. 6 Changes in MeSH and the literature over the years, with respect to Psychiatry and Psychology

PubMed search engine transforms every query, among other things, into a search of the query term together with all of its descendants.

It is the thesis of this paper that the evolution of biomedical terminology reflects the development of biomedical knowledge itself. Figure 6 provides evidence that this is indeed the case.

The figure shows the relationship between the (normalized) growth of psychiatry and psychology articles since 1963 and the degree of change in the F section of MeSH over that same time period. Notice that the peak in the literature in the 1968–

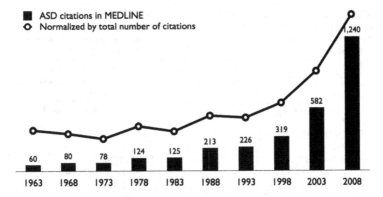

Fig. 7 Actual and normalized growth in the ASD literature

1973 time frame is reflected in extensive changes to MeSH in the subsequent time period 1973–1978. Because we sampled every 5 years, this is only an approximation of when the actual changes to MeSH were made. Nonetheless, the figure indicates that a burst in activity in the literature is soon followed by major activity in the indexing vocabulary itself, thereby reflecting progress in the discipline. Studying this activity in greater detail by looking at the changes in individual terms or clusters of closely related terms reveals a good deal about the state of knowledge itself. Several examples will illustrate this point.

Autism is one of a spectrum of disorders that encompass a range of highly heritable, complex childhood neurodevelopmental disorders, usually diagnosed before the age of three in children from all socioeconomic, ethnic, and cultural backgrounds [37, 38]. ASD's are four times more likely to occur in boys than girls, and family studies show a strong, though not yet well-defined, genetic contribution to this condition.

The 2008 MeSH recognizes four pervasive child development disorders: "Asperger syndrome," "autistic disorder," "Rett syndrome," and "childhood schizophrenia," although this latter term is said to be "an obsolete concept, historically used for childhood mental disorders thought to be a form of schizophrenia." ASD's last throughout a lifetime and are often associated with other comorbid conditions. Table 2 above shows that there has been a growth in the number of terms used to describe the phenomenon and shows which terms coexisted with each other in any given time period. Figure 7 shows the growth of the ASD literature over the years.

The numbers on the graph indicate the number of ASD citations in each time period and the superimposed line normalizes the number according to the overall number of citations in all of MEDLINE during that same time period. Until 1973, there were only two MeSH descriptors to describe the ASD literature. (See Table 2 above.) By 1993 there were five distinct descriptors, but Asperger syndrome was not yet among them. What this tells us is that the descriptor "Autism" actually had a rather different (more diffuse) meaning in 1963 when it only shared the

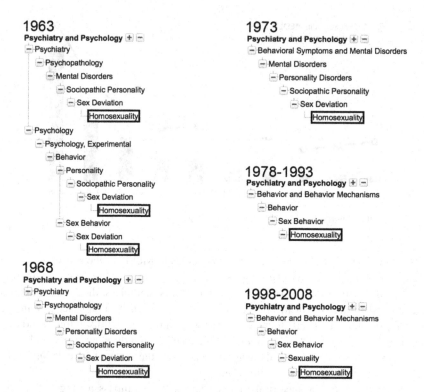

Fig. 8 Placement of the descriptor "Homosexuality" in the MeSH hierarchy over time

ASD conceptual space with "childhood schizophrenia" from the meaning it had in 1993 when there were several other concepts in the same space. That is, "infantile autism," "Rett syndrome," and "pervasive child development disorders" were now also available for indexing. (Interestingly, a number of articles during these various time periods are concerned specifically with differentiating ASD disorders from each other [39, 40].) Any study that looks at the trends in ASD research would do well to recognize these shifts in meaning (and knowledge) over time.

The changes affecting the descriptor "Homosexuality" serve as another interesting example of the reflection in the MeSH terminology of a shift in our understanding of a particular phenomenon. Figure 8 shows the placement of the descriptor "Homosexuality" in the MeSH hierarchy during each of our time periods.

Notice that in the early years "Homosexuality" is a child of "Sex Deviation" in the hierarchy and appears in the "Mental Disorders" subtree. By 1978 "Homosexuality" appears exclusively under the "Behavior" subtree. The significant change in the structural placement of this descriptor can be directly tied to a statement published by the American Psychiatric Association (APA) in late 1973. In this policy document, which resulted in the removal of the term "homosexuality" from the Diagnostic and Statistical Manual of Mental Disorders [41], the APA

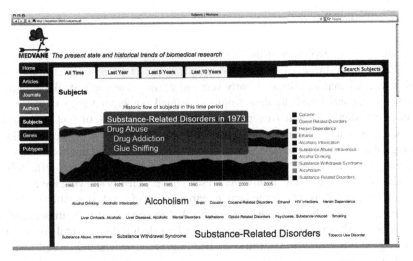

Fig. 9 View of substance-related disorder terms over the years

stated, "The proponents of the view that homosexuality is a normal variant of human sexuality argue for the elimination of any reference to homosexuality in a manual of psychiatric disorders because it is scientifically incorrect, encourages an adversary relationship between psychiatry and the homosexual community, and is misused by some people outside of our profession who wish to deny civil rights to homosexuals" [42].

Our initial hypothesis was that changes in MeSH reflect progress in biomedical research. The work reported here confirms this hypothesis and, therefore, gives us a tool to better detect and analyze trends in specific areas of biomedical research. While it may be tempting to assess research trends over time by counting the number of articles indexed with a specific MeSH descriptor or set of MeSH descriptors, the problem with that approach is that today's MeSH descriptors reflect only the most current view of a particular topic. A more precise view would be one that allows an interested investigator to see exactly which terms were used during which time periods, thereby illuminating the continuing progress of biomedical research and our understanding of specific biomedical phenomena.

Figure 9 shows how this might be done in a system such as Medvane. The figure focuses on the MeSH descriptor "Substance-related Disorders." Recall (from Fig. 4) that this descriptor has undergone quite a few changes over the years. Figure 9 shows the historical trend in the literature about this topic. Note the burst in activity in the early 1970s. Highlighting this area reveals, at a glance, which specific MeSH terms were used during this time period, and, therefore, allows a more refined and specific view of the evolution of the literature about this general concept.

5 Conclusion

Our study of the evolution of the MeSH terminology since its early beginnings in the 1960s to the present day revealed a wide range of interesting terminological and structural changes. Because early versions of MeSH are only available in print form, we reconstructed the data from the printed books, and then only for one section of MeSH, sampled every five years. For the results of our study to be broadly applicable, all earlier versions of MeSH would need to be made available in digital form. With these data in hand, more sophisticated and nuanced studies of the evolution of biomedical knowledge would be possible. The MeSH terminology provides a rich window into biomedical research, very clearly showing the evolution in our understanding of biomedical phenomena. The work described here can be seen as a contribution both to research that is concerned with understanding the conceptual structure and evolution of a knowledge domain and to research whose goal it is to develop optimal methodologies for understanding and tracking ontological change.

Acknowledgements We thank the MeSH staff at the National Library of Medicine for providing us electronic versions of MeSH tree structure files. We thank Malika McCray for data entry and quality control. This work was supported in part by Harvard Catalyst—The Harvard Clinical and Translational Science Center (NIH Grant #1 UL1 RR 025758-01 and financial contributions from Harvard University and participating academic health care centers).

References

1. Library of the Surgeon-General's Office (U.S.): Index-Catalogue of the Library of the Surgeon-General's Office. Government Printing Office, Washington (1880–1961)
2. Guidelines for the construction, format, and management of monolingual thesauri. ANSI/NISO Z39.19-1993
3. Cote, R.G., Jones, P., Apweiler, R., Hermjakob, H.: The Ontology lookup service, a lightweight cross-platform tool for controlled vocabulary queries. BMC Bioinformat. **7**, 97 (2006)
4. Lambrix, P., Tan, H., Jakoniene, V., Str'mb'ck, L.: Biological ontologies. Semantic Web, pp. 85–99. Springer, Berlin (2007)
5. Reed, S., Lenat, D.: Mapping ontologies into cyc. In: Proceedings of the AAAI Conference 2002 Workshop on Ontologies for the Semantic Web. AAAI Press (2002)
6. Gruber, T.: Toward principles for the design of ontologies used for knowledge sharing. Int. J. Human Comput. Stud. **43**, 907–928 (1995)
7. McCray, A.T., Nelson, S.J.: The representation of meaning in the UMLS. Methods Inf. Med. **34**, 193–201 (1995)
8. Stuart, J.N., Douglas, J., Betsy, L.H.: Relationships in medical subject headings. In: Carol, B.A., Rebecca, G. (eds.) Relationships in the Organization of Knowledge, pp. 171–184. Kluwer, New York (2001)
9. MeSH Vocabulary Changes. National Library of Medicine. http://www.nlm.nih.gov/mesh/2009/introduction/intro_voc_change.html. Accessed 28 Feb 2013
10. MEDLINE/PubMed Data maintenance overview. U.S. National Library of Medicine. http://www.nlm.nih.gov/bsd/licensee/medline_maintenance.html. Accessed 28 Feb 2013

11. Medical subject headings: citation maintenance tasks in XML format. U.S. National Library of Medicine. http://www.nlm.nih.gov/mesh/gcmdoc2008.html. Accessed 28 Feb 2013
12. MeSH Memorandum of Understanding. U.S. National Library of Medicine. http://www.nlm.nih.gov/mesh/2009/download/termscon.html. Accessed 28 Feb 2013
13. Chen, C., Paul, R.: Visualizing a knowledge domain's intellectual structure. Computer **34**, 65–71 (2001)
14. Chen, C.: Searching for intellectual turning points: progressive knowledge domain visualization. Proc. Natl. Acad. Sci. USA **101**(Suppl 1), 5303–5310 (2004)
15. Chen, T., Lee, M.: Revealing themes and trends in the knowledge domain's intellectual structure. Lect. Notes Comput. Sci. **4303**, 99–107 (2006)
16. Kleinberg, J.: Temporal dynamics of on-line information streams. Data Stream Management: Processing High-Speed Data Streams. Springer, Berlin (2005)
17. Kontostathis, A., Galitsky, L.M., Pottenger, W.M., Roy, S.: A survey of emerging trend detection in textual data mining. Survey of Text Mining: Clustering. Springer (ISBN 0387955631) (2003)
18. Mane, K.K., Borner, K.: Mapping topics and topic bursts in PNAS. Proc. Natl. Acad. Sci. USA **101**(Suppl 1), 5287–5290 (2004)
19. Roy, S., Gevry, D., Pottenger, W.M.: Methodologies for trend detection in textual data mining. In: Proceedings of the Textmine, vol. 2, pp. 1–12 (2002)
20. Shiffrin, R.M., Borner, K.: Mapping knowledge domains. Proc. Natl. Acad. Sci. USA **101**(Suppl 1), 5183–5185 (2004)
21. White, H.D., Lin, X., Buzydlowski, J.W., Chen, C.: User-controlled mapping of significant literatures. Proc. Natl. Acad. Sci. USA **101**(Suppl 1), 5297–5302 (2004)
22. Noy, N., Kunnatur, S., Klein, M., Musen, M.: Tracking changes during ontology evolution. Lecture Notes in Computer Science, pp. 259–273. Springer, Heidelberg (2004). http://link.springer.com/chapter/10.1007%2F978-3-540-30475-3_19
23. Klein, M., Fensel, D.: Ontology versioning on the Semantic Web. Proceedings of the SWWS, pp. 75–91. Stanford University (2001)
24. Noy, N.F., Musen, M.A.: PROMPTDiff: a fixed point algorithm for comparing ontology versions. In: 18th National Conference on Artificial Intelligence. AAAI-2002. AAAI Press (2002)
25. Klein, M., Fensel, D., Kiryakov, A., Ognyanov, D.: OntoView: comparing and versioning ontologies. ISWC 2002, Sardinia, Italy, 2002
26. Voelkel, M., Groza, T.: SemVersion: an RDF-based ontology versioning system. In: Proceedings of the IADIS international conference WWW/Internet. Stanford University (2006)
27. Euzenat, J., Shvaiko, P.: Ontology Matching. Springer, Berlin (2007)
28. Zhang, S., Bodenreider, O.: Experience in aligning anatomical ontologies. Int. J. Semant. Web Inf. Syst. **3**, 1–26 (2007)
29. Flouris, G., Plexousakis, D.: Handling ontology change: survey and proposal for a future research direction. TR-362, FORTH-ICS (2005)
30. Kang, J., Naughton, J.: On schema matching with opaque column names and data values. In: Proceedings of the 22nd International Conference on Management of Data (SIGMOD), pp. 205–216. ACM (2003). http://dl.acm.org/citation.cfm?doid=872757.872783
31. Specialist NLP Tools. http://specialist.nlm.nih.gov. Accessed 28 Feb 2013
32. McCray, A.T.: The nature of lexical knowledge. Methods Inf. Med. **37**, 353–360 (1998)
33. Unified Medical Language System. http://www.nlm.nih.gov/pubs/factsheets/umls.html. Accessed 28 Feb 2013
34. Resnik, P.: Semantic similarity in a taxonomy: an information-based measure and its application to problems of ambiguity in natural language. J. Artif. Intell. Res. **11**, 95–130 (1999)
35. Wu, Z., Palmer, M.: Verb semantics and lexical selection. In: Proceedings of the 32nd Annual Meeting on Association for Computational Linguistics. Association for Computational Linguistics, Las Cruces, New Mexico (1994)
36. Entrez Utilities. http://www.ncbi.nlm.nih.gov/books/NBK25500/. Accessed 28 Feb 2013

37. Folstein, S.E., Rosen-Sheidley, B.: Genetics of autism: complex aetiology for a heterogeneous disorder. Nat. Rev. Genet. **2**, 943–955 (2001)
38. Fombonne, E.: Epidemiological surveys of autism and other pervasive developmental disorders: an update. J. Autism. Dev. Disord. **33**, 365–382 (2003)
39. Dauner, I., Martin, M.: Autismus Asperger oder Fruehschizoprenie? Zur nosologischen Abgrenzung beider Krankheitsbilder. Padiatr. Padol. **13**, 31–38 (1978)
40. Howlin, P.: Outcome in high-functioning adults with autism with and without early language delays: implications for the differentiation between autism and Asperger syndrome. J. Autism. Dev. Disord. **33**, 3–13 (2003)
41. Diagnostic and Statistical Manual of Mental Disorders, 4th edn. DSM-IV. American Psychiatric Association, Washington, DC (1994)
42. American Psychiatric Association: Homosexuality and Sexual Orientation Disturbance: Proposed Change in DSM-II, 6th printing, p. 44. Position statement (1973)

Crystallizations as a Form of Scientific Semantic Change: The Case of Thermodynamics

C. Ulises Moulines

Abstract When considering the diachronic structure of scientific disciplines, four ideal types of change in the semantic systems constituting them may be distinguished: (1) (normal) evolution; (2) (revolutionary) replacement; (3) embedding (of one conceptual system into another); and (4) (gradual) crystallization of a conceptual framework. In the literature on diachronic philosophy of science, the last type has been unduly neglected in spite of its great significance for understanding the history of science. It consists in a piecemeal but nevertheless fundamental change of the conceptual framework of a discipline during a relatively long period of time. In this paper, an attempt is made to characterize as precisely as possible what the general semantic features of crystallization are and to illustrate them by analysing the gradual emergence of phenomenological thermodynamics in the middle of the nineteenth century, in particular as due to the work of Rudolf J. Clausius between 1850 and 1854. The (formal-semantic) notion of classes of models as understood in the structuralist metatheory of science will be used for that purpose. Three different conceptual nets will formally be distinguished in Clausius's papers corresponding to as many steps in the crystallization of what is now known as (phenomenological) thermodynamics. The reconstruction makes explicit the way Clausius's concepts evolved during the historical period under consideration.

1 Introduction

Scientific disciplines consist, among other things, of particular units usually called "scientific theories". These units are relatively well-defined conceptual systems in the sense that they are built out of a particular set of specific concepts of

C.U. Moulines (✉)
Department of Philosophy, Logic and Philosophy of Science, Ludwig-Maximilians-Universität München, Munich, Germany
e-mail: moulines@lrz.uni-muenchen.de

B.-O. Küppers et al. (eds.), *Evolution of Semantic Systems*,
DOI 10.1007/978-3-642-34997-3_11, © Springer-Verlag Berlin Heidelberg 2013

their own, which they apply in specific ways to a given portion of empirical reality—that portion a particular theory is supposed to deal with. In any given scientific theory, the semantics of its specific concepts and the methodology of its application to an intended range of experience are essentially inter-twinned. Now, scientific theories (and, especially, *empirical* scientific theories) are not static entities[1]: A full grasping of their nature has to be undertaken from a diachronic perspective. Scientific theories, and the conceptual frameworks they are associated with, are like organisms—it makes sense to say of them that they *change*. Since theories are the stuff (or, more cautiously, *one* stuff) scientific disciplines are made of, major changes in the theories that make up a discipline also imply major changes in the discipline itself. An interesting task for the philosophy of science is to determine the different types of change that may occur within a given discipline. Scientific theories are in a perpetual state of change, and so are the disciplines containing them. This is no surprise. Science is constantly in flux. However, a great number of these changes have no deep epistemological and semantical significance, since they don't involve a *conceptual* change—or, at most, the theory's concepts change extensionally but not intensionally. Consider, for example, the quite usual case where a given parameter in a given theory is determined with some more exactness than before (say, by adding some decimals); or consider the case where a theory that has already been applied successfully to a number of systems is subsequently applied to a new system that looks very much like the previous ones. In such cases, there is certainly a change in the theory and it may be quite important for the practice of science or for its technological applications, but it is not the one that we should take seriously as a *genuine conceptual* change. Sure, the theory's concepts change a bit, since their extensions change a bit, but not so their intensions in the sense of their position in the theory's conceptual frame—the interrelationships between the theory's fundamental concepts and the laws they appear in remain the same. These are not changes that should worry the philosophers of science much.

Consider a discipline D (say, physics, or a branch of physics) that consists of the theories $T_1, T_2, \ldots, T_n$. In a first step, we may consider two sorts of changes in D that are significant from an epistemological and semantical point of view. In one case, there may be a significant change in one of the theories constituting D, say in T_i, although T_i remains "the same"—in a sense of "the same" that still has to be explicated. In this case, strictly speaking, there is a conceptual change in T_i but not in D as a whole, since D still consists of the same set $T_1, T_2, \ldots, T_n$. On the other hand, there may be more substantial changes in D. One type of them appears to be the case where T_i is eliminated from D and completely replaced by a T_k that is incompatible with the theory eliminated. Clearly, this is an important change for D itself. A second possibility is that, while T_i is retained, a new theory T_k appears to which T_i has a particular sort of subordinate relationship. Finally, we may consider the case where T_i changes gradually until, after a more or less long process,

---

[1]In the following, by "a (scientific) theory" I always refer only to *empirical* theories. Problems related to the nature of purely mathematical theories lie outside the scope of this paper.

it eventually results in a new T_i', which has some kind of "family resemblance" with the original T_i but is actually a different theory. In all the three kinds of changes just mentioned, the evolution of the discipline itself is accompanied with (or, rather, is conditioned by) more or less drastic conceptual changes in its theories.

All of this is still rather vague and metaphorical. The question is whether we can make these intuitive distinctions between different kinds of changes *formally* precise. I think we can, but only if we don't conceive of theories as just being axiom systems in the way classical philosophy of science did. We have to abandon the statement view of theories and work instead with a non-statement or "modellistic" view. Not the notion of a scientific statement but that of a scientific model is to be regarded as the basic unit when dealing with the analysis of theories in general, and more particularly with the different sorts of conceptual changes that show up in a diachronic perspective on science. This is an insight that has gained momentum in the recent development of philosophy of science among several authors and schools. One such modellistic approach is what is known as the "structuralist view of theories", or "structuralism" for short, set forth by Joseph D. Sneed, Wolfgang Stegmüller, Wolfgang Balzer and myself, among others. Here, I'm going to use *some* of the methodological instruments of structuralism to deal with the questions of conceptual change at stake, and in particular with the last kind of conceptual evolution listed above. For those readers who are not well-acquainted with structuralist methodology in general, a brief summary of the notions required for our present purposes is laid out in the next section.[2]

2   The Essentials of Structuralism

The first thing to notice is that, according to structuralism, the term "theory" is polysemic: Its ordinary use denotes in fact different kinds of entities, which differ according to their levels of aggregation. For the present purposes, it suffices to consider two of the formal concepts that structuralism has proposed as explications of the intuitive notion of a theory: *theory-elements* and *theory-nets*. A theory-net is a hierarchically organized array of theory-elements with a "common basis" and a common conceptual framework. The "essential identity" of a theory-element is, in turn, given by two constituents: a homogeneous class of models and a domain of intended applications described by partial structures of the models (their relative non-theoretical parts). (Besides these two constituents, according to structuralism, a "normal" theory-element is also constituted by other, more complex model-theoretic structures; nevertheless, for the purposes of the present paper, we may forget about them.)

---

[2]The standard exposition of the structuralist metatheory (with several examples of application) is [1].

Now according to the model-theoretic tradition going back to the works of Alfred Tarski and Patrick Suppes, a model is a structure of the form

$$\langle D_1, \ldots, D_m \, ; \, A_1, \ldots, A_n \, ; \, R_1, \ldots, R_p \, ; \, f_1, \ldots, f_q \rangle, \tag{1}$$

where the D_i are the theory's basic domains (its "ontology"), the A_i are auxiliary sets of mathematical entities (usually numbers or vectors), the R_i are relations on (some of) the basic sets, and the f_i are (metrical) functions on (some of) the basic and auxiliary sets.

It is not very important how we determine these models—whether by thoroughly formal means, or semi-formally, or in ordinary language. But it seems that the most practical method is to define the class of a theory's models by means of the Suppesian method of defining a so-called *set-theoretical predicate*, i.e. a predicate defined solely by means of the tools of naive or semi-formal set theory. (We'll see several examples of such predicates below.)

A homogeneous class of models, **M**, is a class of structures satisfying all the same set-theoretical predicate. In empirical theories, to this class of models a domain of intended interpretations, **I**, is associated: This is a class of substructures of *some* of the elements of **M**, containing (almost) all the D_i and A_i, but only some of the R_i and f_i—those that are non-theoretical with respect to the theory in question, i.e. those whose values may be determined without presupposing the substantial laws used to define the set-theoretical predicate which, in turn, determines **M**. The other relations and functions, if any, are called "relatively theoretical", or "T-theoretical". The domain of intended applications of a given theory represents the physical systems to which the theory is supposed to apply, where it has to be tested, or in other words, its empirical basis (in a relative sense).

Though the distinction between T-theoretical and T-non-theoretical notions is an important constituent of the structuralist metatheory, its application to concrete case studies is usually a quite intricate issue, which involves a lengthy and careful discussion. For lack of space, I'll not go into this issue in the case study analysed in this paper.

Thus, a theory-element is (in this highly simplified version of structuralist methodology) an ordered pair $\langle \mathbf{M}, \mathbf{I} \rangle$

A theory-net is a finite class of theory-elements ordered by a relation of *specialization* and having a first element. This first element is what may be called "the *basic theory-element*", $\mathbf{T_0}$ ($= \langle \mathbf{M_0}, \mathbf{T_0} \rangle$). It is the one whose models are determined only by the fundamental law(s) of the theory in question. Taken in isolation, the fundamental law(s) in $\mathbf{T_0}$ usually have an empirically almost vacuous nature—they are "guiding principles" rather than statements of fact. But for this very reason, they constitute the theory's core. All other theory-elements $\mathbf{T_i}$ of the net come out of $\mathbf{T_0}$ by a process of specialization, i.e. by adding more special (and more contentful) laws and conditions to the fundamental law(s), which frequently involve the addition of new, more specific concepts, and by restricting the range of intended applications to some specified systems. The relation of specialization between theory-elements (let's symbolize it by "σ") may be constructed out of

Fig. 1 A theory-net

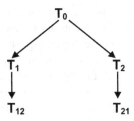

the set-theoretic relationships of "inclusion" and of "partial structure". I'll skip the technical details here. Typically, a theory-net has the form of an inverted tree (Fig. 1), where $T_1 \sigma T_0$, $T_{12} \sigma T_1$, and so on. Since the specialization relation is transitive, all theory-elements of a net are, directly or indirectly, specializations of the basic (schematic) theory-element T_0.

Accordingly, the presystematic, intuitive notion of a scientific theory either corresponds to what we call a "theory-element" or else a "theory-net" consisting of several, often very many, different theory-elements. They are all connected to the "top element" by the relation of specialization. Notice that, for this reason, one may say that the conceptual framework of a theory-net is essentially homogeneous in the sense that all the fundamental concepts of the several specialized theory-elements also appear in the basic theory-element. But the concrete semantic interpretation of a given concept depends not only on the theory's fundamental law(s) but, to some extent, also on the more special laws where it appears and on the more particular intended applications the specialization is conceived for. In this way, "minor" changes in secondary theory-elements imply to some extent equally "minor" changes in the overall conceptual system of the theory-net. It is from this model-theoretic perspective on the structure of theory-nets that we can deal now with the problem of conceptual change in a quite precise way.

3 Four Types of Diachronic Processes in a Scientific Discipline

In the introductory section, I've already suggested different kinds of scientific change that appear to be accompanied with more or less substantial conceptual changes. Let's try now to be more precise about them. The structuralist explication of theories as nets of theory-elements will help us in this endeavor. As a matter of empirical (historical) fact, it is assumed that at least four *main* types of diachronic processes, which involve some kind of conceptual change, may be distinguished in the history of science. All of them can be, and have been, formally represented within the structuralist framework. These are:

– *Evolution*: A theory-net evolves in time, adding or suppressing some specialized theory-elements but without losing its "essential identity", which is given by the basic element T_0. This kind of process roughly corresponds to what Thomas S.

Kuhn called "normal science" and Imre Lakatos "a research program". Example: the evolution of Newtonian mechanics from the end of the seventeenth century to the beginning of the twentieth century.[3] In this kind of change, the "core semantics" of the fundamental concepts appearing in the conceptual framework of T_0 is preserved, but there appear some minor semantic changes due to the fact that the concepts get more concrete interpretations in different ways in the theory-elements that are specializations of T_0. A further conceptual change may possibly come from the appearance or disappearance of secondary concepts ("parameters") in the specializations that come and go.

- *Replacement*: A theory, i.e. a theory-net, aimed at systematizing a given domain of intended applications is completely replaced by another theory-net (with a very different conceptual framework) aimed at more or less the same domain of intended applications. Formally, the older basic theory-element $\langle M_0, I_0 \rangle$ gets replaced by a new basic theory-element $\langle M_0^\star, I_0^\star \rangle$, where M_0 and $M_0^\star$ are very different, while I_0 and $I_0^\star$ overlap to a substantial degree. This corresponds to one sense of what Kuhn has called a "scientific revolution". Example: the replacement of phlogiston theory by the oxidation theory.[4] This is the more extreme case of conceptual change within a discipline, though (*pace* Kuhn) we may not speak of total semantic incommensurability since the interpretation of the concepts describing I_0 and $I_0^\star$ cannot become completely disparate (unless one wants to claim that there is absolutely no way to compare the relative merits and demerits of the respective theory-nets—a claim that Kuhn himself apparently didn't want to make and that would be historically implausible anyway).

- *Embedding*: The models of a previous theory-net are (approximatively, and perhaps not completely) embedded into the models of a more complex theory. "Embedding" is to be understood here as meaning that the models of the first theory (at least a substantial part of them) become *partial structures* of the models of the second theory. This roughly corresponds to another sense of Kuhnian "revolutions". Example: the (approximative) embedding of Keplerian planetary theory into Newtonian mechanics.[5]

- *Crystallization*: After the breakdown of a previous theory-net, and through a long and gradual process, the models of a new theory are constructed in a piecemeal fashion, with many intermediate, fragmentary states before a fully developed new theory-net appears. Such a process is neither "normal" nor "revolutionary" in anything like Kuhn's sense, and I know of no other philosophers of science who have dealt with this kind of process (with the possible exception of Yehouda Elkana [6]). But I contend that it is quite frequent in the history of science. Crystallization is much more complex than the previously mentioned types of diachronic structures; it contains some elements of the other types but it also has its own features. The slow emergence of phenomenological thermodynamics

---

[3] The detailed structuralist reconstruction of this example will be found in [1, Ch. V.].

[4] For a structuralist reconstruction of this example see [2].

[5] This example is dealt with in [1, Ch. VII].

after the breakdown of the caloric theory seems to be a clear case in point. In this process, the work of Clausius analysed in this paper played a crucial role (though it was only a part of the whole process). In the crystallization of a new theory, some of the old concepts are completely given up, but some others are re-interpreted in a new fashion (according to new fundamental laws and in view of new intended applications), and also some totally new concepts are introduced.

In the next sections, a portion of the crystallization of phenomenological thermodynamics in the nineteenth century is reconstructed according to the general scheme of evolving theory-nets. This portion of the historical process will be reconstructed as three interrelated but essentially different theory-nets developed by Clausius, and we'll see how Clausius's conceptual system gradually evolved.

4   The Historical Data

The "paradigm" for the theory of heat from the end of the eighteenth century until the mid-1820s was the caloric theory developed by Lavoisier, Laplace, Biot, and others. This was a quite successful theory until it began to agonize in the late 1820s due to "internal" (conceptual) as well as "external" (application-related) "anomalies". A new "paradigm" for the theory of heat, i.e. a full-fledged theory which was fully satisfactory both from the conceptual-mathematical and from the empirical point of view, was to appear only much later, in 1876, with J. Willard Gibbs's *On the Equilibrium of Heterogeneous Substances* (first part) [7]. There we have for the first time what we now call "phenomenological thermodynamics" in its presently acknowledged form, almost indistinguishable from what is laid out in contemporary textbooks on thermodynamics. In the meantime, that is, between the 1820s and the 1870s we are confronted with a long and chaotic period of partial attempts at the construction of a fully satisfactory theory of heat, which were partially successful but which also partially failed, with very different theoretical concepts and principles. In this long process, the names of Carnot, Clapeyron, Regnault, Mayer, Helmholtz, Joule, Kelvin, Rankine, and above all Clausius, played an important role. These scientists were working from different standpoints and obtaining different, sometimes incompatible results. The culmination of this process was Gibbsian thermodynamics.[6] Clausius's work was an essential part of this process, itself a fragmentary crystallization within the general crystallization process. This fragment is what is to be reconstructed in the next pages.

In the 1850s, Clausius published three papers on the theory of heat: *"Über die bewegende Kraft der Wärme"* (1850), *"Über eine veränderte Form des zweiten Hauptsatzes der mechanischen Wärmetheorie"* (1854), and *"Über die Art der Bewegung, die wir Wärme nennen"* (1857) [3–5]. The last paper amounts to the

---

[6]For the history of the emergence of Gibbsian thermodynamics see [8].

first consistent exposition of the kinetic theory of gases, and in this sense, it is the forerunner of statistical mechanics; as such, it is of no concern to us here, since we are dealing with the development of (phenomenological) thermodynamics—which is what Clausius, and others, called then the *"mechanische Wärmetheorie"*—the "mechanical theory of heat". So, let's restrict our attention to the papers of 1850 and 1854. The 1850 essay [3] contains one theory, while the 1854 [4] paper contains two theories, which presumably were conceived at different times. Each one of these theories may consistently be represented as a theory-net in our sense. To abbreviate, we'll denote the 1850 theory(-net) by "**Cl-1**", while the two theories (i.e. theory-nets) laid out in the 1854 paper will be denoted by "**Cl-2.1**" and "**Cl-2.2**", respectively.

5   The Basic Theory-Element of Cl-1

Clausius's paper of 1850 [3] is divided into two parts, respectively, entitled: "I. Consequences of the principle of the equivalence of heat and work", and "II. Consequences of Carnot's principle in connection with the former".[7] Given the form of their presentation, one could be misled in thinking that Clausius lays out *two* different fundamental principles: the principle of the heat/work equivalence, and the hypothesis that Clausius considers to be the core of the so-called "Carnot principle", namely that "whenever work produces heat while not producing a permanent transformation in the state of the working body, a certain amount of heat goes from the hot body to a cold one" [3, p. 4]. Now, this interpretation, which is still to be found in contemporary expositions, seems to be misleading, from the point of view both of Clausius's original text and of a coherent formal reconstruction of them. In fact, Clausius himself stresses several times that both principles are "intimately linked". It is more adequate, as we shall see, to interpret both "principles" as two specifications of a more general and fundamental principle that Clausius takes implicitly as his starting point.

To find out how the models of **Cl-1** look like, two preliminary questions have to be dealt with: First, what is the theory's "basic ontology", that is, what kinds of objects are assumed, and second, what relations and functions defined over those objects does one need to formulate the theory's laws.

1. Clausius speaks once and again of *bodies*, that is, of concrete, spatio-temporally located entities that may be in physical contact and/or have component parts. In some passages of his paper, he also has to speak about a more abstract kind of entities: *chemical substances* to which the concrete bodies may belong; but this he does only for some specializations of his theory. This means that his basic

---

[7]Here, as in the rest of the paper, I translate as truthfully as I can from the original German text. The page numbers indicated in the quotations are those of the original essay.

"ontological commitment" is only with bodies, whereas the "commitment" with *kinds* of bodies only occurs in some of the theory's specializations (in particular, for the treatment of ideal gases). Also, Clausius speaks of bodies as being in a certain *state*, or as changing their state. Consequently, the basic sets of each model of **Cl-1** will be:

- A finite set of bodies: $K = \{k_1, k_2, \dots\}$.
- A continuous sequence of states: $Z \cong \Re^{+}.$[8]

2. As for the relational concepts, Clausius needs a certain number of primitive magnitudes to formulate his laws. From a formal point of view, the problem with **Cl-1** is that this theory deals with *two* quite different kinds of intended applications: ideal gases and steam at a maximum density. The special laws relevant for one and the other type of application contain indeed some common concepts, but other concepts are specific either of ideal gases or of steam. Formally, this means that in the theory-net representing **Cl-1**, the models of the basic element will contain some concepts common to all theory-elements of the net, while in the specializations some further concepts have to be added as primitive notions—the models becoming thereby more complex structures.

The magnitudes common to all elements of the theory-net are:

A (a universal constant settling the heat/work equivalence)[9]
P (pressure)
V (volume)
t ("empirical" temperature)
Q ("heat")
W ("work")

Comment: There is no doubt that Q and W are basic concepts in **Cl-1**—whatever usual textbooks and historical expositions, according to which "heat" was not a fundamental notion for Clausius anymore, might claim. Indeed, Clausius retakes these notions from the previous literature, essentially from the caloric tradition, but at the same time he is about to give them a fundamentally new interpretation. He substantially departs from previous approaches to heat and work (essentially those of the caloric theory), especially by his explicit denial that Q is a conservative magnitude and by making it "equivalent" to W. In fact, the discussion at the beginning of his essay seems to make clear that a new interpretation of Q and W is needed, whereby these notions only make sense if one accepts the "principle of equivalence" between Q and W.

---

[8]One needs a continuous sequence of states as arguments of differentiable magnitudes. Clausius assumes throughout his text without further ado that most thermodynamic magnitudes are differentiable functions.

[9]Clausius assumes that, according to Joule's experiments, the value of A should be approximately 1/421. But A's concrete value is irrelevant for the formal reconstruction of the theory.

The specific concepts to be *added* in order to formulate the specializations are, respectively,

1. For the ideal gases:

> Γ (a finite set of gaseous substances—this notion is implicit in the original text)
> R (a parameter specific of each gaseous substance)
> c_V (specific heat at constant volume)
> c_P (specific heat at constant pressure)
> U (an "anonymous" or "arbitrary" function—"willkürlich" is the German word Clausius employs here)[10]
> a (a constant specific of ideal gases)[11]

Comment: In the work of Clausius, R is not a universal constant; it is rather a parameter depending on the substance's specific weight; therefore, it is not identical with our universal constant "R" though it is related to it. As for c_V and c_P, these were notions already used in the last phase of the caloric tradition (e.g. by Clapeyron); Clausius just imports them in its own theory. U is a complete new invention of Clausius; it is undoubtedly what later will be called "internal energy", appearing in this text for the first time in the history of thermodynamics. As for a, it is obviously related to the value of the absolute temperature; but Clausius still doesn't have this notion in his baggage in 1850.

2. For steam at a density maximum:

> $\sum$ (a finite set of substances with a liquid and a steam phase—implicit in the original text)
> s (steam volume at a maximum density)
> σ (liquid volume at a maximum density)
> m (mass of liquid transformed into steam)
> c (the liquid's specific heat)
> h (another "anonymous" parameter)

Comment: The magnitudes s, σ, m, and c had already been amply used in the previous literature, especially by engineers dealing with the optimization of locomotives. The magnitude h is another conceptual invention of Clausius required to formulate certain differential equations for steam processes, of which Clausius only says that it should depend on the steam's temperature.

---

[10]A further interesting aspect of U should be noted: Clausius explicitly avows that he introduces this magnitude in order to formulate a differential equation that will allow for a derivation of the ideal gas law, U making sense only in the context of the equation at stake. In other words, U appears to be Cl-1-theoretical. The same goes for the function h introduced later on—see below.

[11]Clausius writes at the end of his paper that experimental results suggest that the actual value of a should lie "around 273"; but, again, the concrete value is unimportant for the formal reconstruction.

Having settled the conceptual framework of **Cl-1**, the next step is to provide an adequate reconstruction of the fundamental law(s) of the basic theory-element of **Cl-1**. A general metatheoretical hypothesis of structuralism is that, in theories having attained a certain degree of unity and structural complexity, there will be *just one* fundamental law, playing the role of a guiding principle and having (almost) no empirical content when taken in isolation. At first sight, **Cl-1** seems to contravene this hypothesis, since Clausius explicitly states two axioms: the Q/W-equivalence and "Carnot's Principle". However, I'll argue that appearances mislead us here and that the metatheoretical just-one-fundamental-law hypothesis continues to be valid in the present case.

Clausius formulates the Q/W-equivalence principle in the first part of his article [3, p. 17]:

the heat consumed divided by the work produced $= A$.

As for the Carnot Principle, which is relevant for the second part of the essay, Clausius states it in a purely informal way:

to the production of work as equivalent there corresponds a mere transfer of heat from a hot body to a cold one [3, p. 30].

It is striking that, when Clausius goes on to formulate the Carnot Principle in the second part of his article, he *presupposes* that there is an equivalence of work and heat, that is, he presupposes the first principle. And, as already remarked, there are other passages in the article where he emphasizes that both principles are "intimately connected". Therefore, we are confronted with two possible interpretations: (a) the equivalence principle is the theory's only fundamental law and the Carnot Principle is a *specialization* thereof in the structuralist sense; or (b) both principles constitute in fact *two* aspects of one and the same (implicit) guiding principle that is to be formulated synthetically. Both interpretations are coherent with the rest of the original text. Nevertheless, I favor the second one since it is easy to make it explicit and, moreover, it seems to me to be more in accordance with Clausius's original intentions. Indeed, if we admit that Q and W are differentiable functions (as Clausius himself presupposes all the time), then it is immediate to state both principles for any body k being in the state z as:

Equivalence Principle:

$$\frac{dQ(k,z)}{dW(k,z)} = A \in \mathfrak{R}; \tag{2}$$

Carnot's Principle:

$$dW(k,z) > 0 \rightarrow dQ(k,z) < 0. \tag{3}$$

Then, it is immediate to formulate a synthetic version of both preceding formulae:

$$dQ(k,z) = -A \cdot dW(k,z), \text{ where } A > 0.^{12} \tag{4}$$

This is the fundamental law of **Cl-1**. It is easy to see that, as is usual in such kinds of laws, this formula by itself is almost devoid of empirical content since Q and W, taken in isolation, cannot be concretely determined (they are **Cl-1**-theoretical). They have to be put in some kind of testable relationship with the other primitive magnitudes of the theory: P, V, t. And, in fact, in his argumentation, Clausius always presupposes that Q and W are dependent on these magnitudes at least. Speaking in more formal terms, it may be said that Q and W are *functionals* of P, V, t, that is, there are functions f^Q and f^W (which are functions of functions) expressing the general form of the dependence of Q and W on P, V, t.

The preceding considerations allow now for a formal definition of the models of the basic theory-element of the net **Cl-1**. For more perspicuity, let's skip out the explicit indication of the formal features of each single component of the models (i.e. whether it is a finite or infinite set, a differentiable function, and so on) as well as the arguments of the magnitudes (clearly the latter are always functions from physical bodies and their states into real numbers), whenever there is no danger of confusion.

Cl-1: $x \in \mathbf{M_0[Cl\text{-}1]}$ iff: $\exists K, Z, P, V, t, Q, W, A, f^Q, f^W$, such that

(0) $x = \langle K, Z, \Re, A, P, V, t, Q, W \rangle$
(1) $Q = f^Q(P, V, t, \dots)$
(2) $W = f^W(P, V, t, \dots)$
(3) $A > 0$
(4) $dQ = -A \cdot dW$

We could synthesize conditions (1)–(4) into one "big" formula:

$$(\mathbf{FL_{Cl\text{-}1}}) \ df^Q(P, V, t, \dots) = -A \ df^W(P, V, t, \dots), \text{ for } A > 0. \tag{5}$$

This is the theory's fundamental law.

What about the intended applications of this basic theory-element? Clausius himself clarifies this point when he writes: "[in this work] we are going to restrict our attention to permanent [i.e. ideal] gases and to steam at a maximum of its density" (p. 11). Therefore, the basic domain of intended applications of **Cl-1**, **I[CI-1]**, is

I[CI-1] = {ideal gases ; steam at a maximum density}.

Thus, the net's basic theory-element is: $\langle \mathbf{M_0[Cl\text{-}1]}; \mathbf{I[CI\text{-}1]} \rangle$, as defined.

---

[12]It is not clear in the original text whether Clausius wanted to conceive A as a positive or a negative value. In the first case, of course, he should have worded his "Equivalence Principle" slightly differently. But this is a minor point.

6   The Specializations of Cl-1

Two main specialization "branches" may be identified in the tree representing
the theory-net of **Cl-1**: the one corresponding to the study of ideal gases and
the one aimed at steam at a maximum density. The corresponding specializations
are obtained by specifying, in different ways, the functionals f^Q and f^W of the
fundamental law.

For the *ideal gases*, Clausius takes up some previous results to propose the
following specifications:

$$f^W = \frac{R\,dV\,dt}{V} \quad \text{(p. 15)};\tag{6}$$

$$f^Q = \left(\frac{d}{dt}\left(\frac{dQ}{dV}\right)\right) - \frac{d}{dV}\left(\frac{dQ}{dt}\right)dVdt \quad \text{(p. 17)}.\tag{7}$$

As Clausius himself recognizes, the second specification is somewhat problematic
since the notion of a "heat differential" is itself physically and mathematically
dubious. However, by combining it with the less problematic determination of
f^W, Clausius comes to the conclusion that one can establish a more plausible
differential equation if we *postulate* that there must exist an "arbitrary" function U
that allows for a mathematically correct formulation. (I have already remarked that
this "U" is "our" *internal energy*.) Therefore, here we have an example of a brand-
new theoretical concept introduced for the purpose of giving the theory a coherent
conceptual framework; this new notion only makes sense if we already assume the
validity of the theory's laws, and in particular the following specialization of the
fundamental law:

$$dQ = dU + A \cdot \frac{a+t}{V} \cdot dV.\tag{8}$$

This is the most general specialization for ideal gases. Using the additional primitive
concepts we have already indicated above (U, R, c_V, c_P, a, and the implicit notion
of the set of gaseous substances, Γ) we come to the following definition of the
models of this specialization:

G $: x \in \mathbf{M_G}$ iff: $\exists\, K, Z, \Gamma, V, t, Q, W, U, R, c_V, c_P, A, a, f^Q, f^W$, such that

(0) $x = \langle K, Z, \Gamma, \Re, A, a, P, V, t, Q, W, U, R, c_V, c_P \rangle$
(1) $\langle K, Z, \Re, A, P, V, t, Q, W \rangle \in \mathbf{M_0[Cl\text{-}1]}$
(2) $dQ = dU + A \cdot R \cdot \frac{a+t}{V} \cdot dV$

For Clausius, the second term of the sum in (2) represents the "external work" done
by the gas. One obtains a specialization of this specialization by just assuming that
this work has simply the form "$p\,dV$". In such a case, by a simple calculus we obtain
the law of ideal gases. We may call "**IG**" this specialization of the specialization **G**:

IG : $x \in \mathbf{M_{IG}}$ iff: $\exists K, Z, \Gamma, P, V, t, Q, W, U, A, a, f^Q, f^W$, such that

(0) $x = \langle K, Z, \Gamma, \Re, A, a, P, V, t, Q, W, U \rangle \in \mathbf{M[G]}$

(1) $A \cdot R \cdot \frac{a+t}{V} \cdot dV = p(k,z)\, dV(k,z)$

A further specialization on this same line consists in what Clausius calls an "auxiliary hypothesis" [3, p. 24] about the *molecular* constitution of ideal gases: Such gases, when they receive heat, don't do an "internal work"; consequently, all the work done is "exterior" and U is solely a function of temperature. By abbreviating a little, we may define this specialization as follows:

$$\mathbf{Mol} : \quad x \in \mathbf{M[IG]} \quad \wedge \quad \exists \emptyset (U = \emptyset(t)). \tag{9}$$

Subsequently, Clausius makes an additional assumption: "Probably", in this case $\emptyset$ is a simple function determined only by c_V, which allows, in turn, for a determination of c_P:

$$\mathbf{Mol'} : \quad x \in \mathbf{M[Mol']} \quad \wedge \quad U = c_V \cdot t \quad \wedge \quad c_P - c_V = A \cdot R. \tag{10}$$

In the models of $\mathbf{M[Mol']}$ one obtains as a corollary Poisson's law:

$$\forall \gamma \in \Gamma : \; c_p(\gamma)/c_V(\gamma) \text{ is a constant.} \tag{11}$$

The other branch of specializations in the net corresponds to *steam at a maximum density*. The specifications of the heat and work functionals are quite different from those of the ideal gases. And the additional primitive concepts needed (as has been indicated above) are also different. (Analogously to the case of ideal gases, we need the implicit set of those substances that are a particular mixture of a liquid and a vaporous phase, Π.)

V : $x \in \mathbf{M_V}$ iff: $\exists K, Z, \Pi, P, V, t, Q, W, s, \sigma, m, c, h, A$, such that

(0) $x = \langle K, Z, \Pi, \Re, A, P, V, t, Q, W, s, \sigma, m, c.h \rangle$

(1) $\langle K, Z, \Re, A, P, V, t, Q, W \rangle \in \mathbf{M_0[Cl\text{-}1]}$

(2) $dQ = df^Q(P, V, t, \ldots) = dV/dt + (c - h) \cdot dm \cdot dt \quad \wedge$
    $dW = df^W(P, V, t, \ldots) = (s - \sigma) \cdot dP \cdot dm$

Subsequently, Clausius introduces an additional hypothesis suggested by the experimental results of Regnault and Pambour, namely that h always is negative.

$$\mathbf{V'} : x \in \mathbf{M[V]} \quad \wedge \quad \forall k \in K \quad \forall z \in Z : h(k,z) < 0. \tag{12}$$

We are now in a position to build up the graph representing the net **Cl-1** (Fig. 2), where the arrows represent successive specializations.

Fig. 2 The first theory-net of
Clausius's thermodynamics

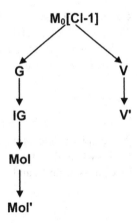

7 The Basic Theory-Element of Clausius-2.1

In his 1854 paper, Clausius [4] makes a new attempt at establishing phenomenological thermodynamics on a well-founded basis. This attempt is more restricted
than **Cl-1**: It almost exclusively deals with cycles ("*Kreisprozesse*"). He defines
a cycle as a "transformation" ("*Verwandlung*") through which a body comes
back to its original state. Clearly, this new concept is inspired by Carnot's (and
Clapeyron's) pioneering theoretical work on the steam engine. However, it is only
in Clausius's second theory that it appears as an "official" scientific notion of the
new discipline of thermodynamics. Another important difference with respect to the
first article is that Clausius presents now the "two principles of thermodynamics"
in a completely independent manner, as if they would correspond to two different
theories. The article is divided into two sharply separated parts with quite different
titles, the second part being much longer and better articulated. Even by the style of
exposition, one gets the impression that the two parts were written at two different
periods of the development of Clausius's thought. The two principles now get the
wording:

I. "Principle of the equivalence of heat and work"
II. "Principle of the equivalence of transformations".

Formally, there is no other way but to reconstruct the two parts as two different
theory-nets, which we will call "**Cl-2.1**" and "**Cl-2.2**", respectively. Let's analyse
first **Cl-2.1**. Other than in the 1850 article, Clausius now states the heat/work
equivalence principle by immediately introducing the "anonymous" function U as
a basic concept, and he makes the following symptomatic remark about it: "U [is
a magnitude] that we cannot specify at present, but of which we know at least that
it is completely determined by the initial and the final state of the body" (p. 130).
Consequently, he formulates the fundamental law of **Cl-2.1** as follows:

$$Q = U + A \cdot W. \tag{13}$$

Comment: The symbol "W" employed here by Clausius doesn't refer now to the "total work", as in **Cl-1**, but only to the "external work".

The first specialization considered refers to cycles. For such cases, Clausius writes: $U = 0$. Presumably, what he wants to express is what we now would write as

$$\oint dU = 0 \tag{14}$$

A second specialization considered in this part of the paper is devoted to those bodies where the pressure is the same on all points (the intended applications being here gases, liquids, and some solids). For this, Clausius proposes the equation

$$dQ = dU + A \cdot p \cdot dV. \tag{15}$$

Towards the end of this first part of the essay, Clausius suggests that one could consider further specializations of this specialization "by applying this formula to certain kinds of bodies. For the two most important cases, namely for ideal gases and for steam at a maximum density, I have developed this more special application in my precedent essay" (p. 133). Obviously, he is referring to the net **Cl-1**.

Let's undertake now the structuralist reconstruction of **Cl-2.1**. Clearly, the conceptual framework is essentially the same as for the models of the basic theory-element of **Cl-1** but for the additional primitive concept "U". That is, the basic models have the form $\langle K, Z, \Re, P, V, t, Q, W, U \rangle$. Formally, this means that the models of $\mathbf{M_0[Cl\text{-}1]}$ become now substructures of the models of $\mathbf{M_0[Cl\text{-}2.1]}$. $\mathbf{M_0[Cl\text{-}2.1]}$ is determined by the same conditions as $\mathbf{M_0[Cl\text{-}1]}$, except for the introduction of an additional "functional" to express the dependence of U with respect to V and t [4, p. 131] and for the new form of the fundamental law:

Cl-2.1: $x \in \mathbf{M_0[Cl\text{-}2.1]}$ iff: $\exists\, K, Z, P, V, t, Q, W, A, f^Q, f^W, f^U$, such that

(0) $x = \langle K, Z, \Re, A, P, V, t, Q, W, U \rangle$
(1) $Q = f^Q(P, V, t, \dots)$
(2) $W = f^W(P, V, t, \dots)$
(3) $A > 0$
(4) $U = f^U(V, t)$
(5) $Q = U + A \cdot W$

8 Specializations of Cl-2.1

The first specialization of this theory-net deals with cycles. Intuitively, a cycle is a process, that is, in our terms, a sequence Z of states with a first element (the "initial state") and a last element (the "final state") such that the values of all fundamental magnitudes are the same in the two states. We could define this notion in completely formal terms within our model-theoretic framework, but since this involves some technicalities, for reasons of brevity, let's just presuppose that the notion of a cycle has already been defined. Then, the specialization looks like this:

$\mathbf{C} : x \in \mathbf{M[C]}$ iff: $\exists\, K, Z, P, V, t, Q, W, A, U$ such that

(0) $x = \langle K, Z, \Re, A, P, V, t, Q, W, U \rangle$ and $x \in \mathbf{M_0[Cl\text{-}2.1]}$
(1) Z is a cycle
(2) $\forall k \in K : \oint dU(k, z) \cdot dz = 0$

As remarked, the other specialization Clausius considers in this part of the article is completely independent of the previous one: In it, it is assumed that W takes a particularly simple form: $P \cdot dV$. We may call it the "uniform pressure specialization", $\mathbf{UP}$.

$\mathbf{UP} : x \in \mathbf{M[UP]}$ iff: $\exists\, K, Z, P, V, t, Q, W, A, U$ such that

(0) $x = \langle K, Z, \Re, A, P, V, t, Q, W, U \rangle$ and $x \in \mathbf{M_0[Cl\text{-}2.1]}$
(1) $dQ = dU + A \cdot P \cdot dV$

As Clausius himself indicates, from this latter specialization one could obtain the law of ideal gases as well as the laws for steam systems. This means that we can *embed* one part of the previous theory-net $\mathbf{Cl\text{-}1}$ into the new net $\mathbf{Cl\text{-}2.1}$. This is a typical move in a crystallization process.

The graph of the theory-net $\mathbf{Cl\text{-}2.1}$ is accordingly (Fig. 3).

9   The Basic Theory-Element of Clausius-2.2

As already noted, the theory-net $\mathbf{Cl\text{-}2.2}$ is exclusively devoted to the most general treatment of the "Second Principle" as completely independent of the "First Principle". And its intended domain of applications are solely cycles.

To begin with, Clausius provides an intuitive formulation of the "Second Principle" that he considers to be more general than the one attributed to previous researchers (notably Carnot and Clapeyron): "Heat can never go from a cold to a hotter body unless another transformation takes place that depends on the first one" [4, p. 134]. Clausius sets himself the task of finding a mathematically appropriate formulation of this principle. The first step consists in introducing the new notion of "equivalence value" ("*Aequivalenzwerth*") of a transformation (of heat into work or conversely). His proposal is

Equivalence value of W : $f(t) =_{df} Q \cdot \dfrac{1}{T(t)}$.

Here, there is a new "anonymous" function T, of which Clausius says: "T is an unknown function of the temperature appearing in the equivalence values" [4, p. 143]. This is for us again an example of a new theoretical concept introduced for the purpose of providing a coherent framework to the theory. Only at the end of the paper, Clausius *suggests* that T may simply be regarded as the *absolute temperature*.

Fig. 3 The second theory-net
of Clausius's thermodynamics

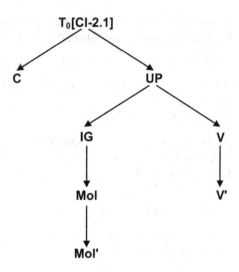

Before the new formulation of the "Second Principle", Clausius establishes a distinction between *reversible* ("*umkehrbar*") and "irreversible" ("*unumkehrbar*") cycles. The text doesn't contain any precise definition of these terms; apparently, the reader is supposed to have a clear intuition of these notions. Anyway, after a long and cumbersome argument, Clausius comes to the conclusion that the formal condition for reversible cycles is

$$\oint \frac{dQ}{T} = 0. \tag{16}$$

A specialization of this condition is Clapeyron's equation

$$\frac{dT}{dt} \bigg/ T = \frac{A}{C}, \tag{17}$$

where C is the so-called "Carnot function".

For irreversible cycles, Clausius concludes that the condition should be

$$\oint \frac{dQ}{T} > 0. \tag{18}$$

If we admit the "additional hypothesis" ("*Nebenannahme*"), which Clausius takes from Regnault, according to which "an ideal gas, when expanding at constant temperature, absorbs only the quantity of heat needed for the external work done" [4, p. 153], we obtain:

$$\frac{dQ}{dV} = A \cdot P \tag{19}$$

and if we add the ideal gas law, we get as a *theorem* $T = a + t$, where a "presumably" is 273. That is why we can consider T to be the absolute temperature.

Clausius never explicitly formulates the fundamental law common to both reversible and irreversible cycles but it is clear that this can be nothing but

$$\oint \frac{dQ}{T} \geq 0. \tag{20}$$

Let's proceed now to the structuralist representation of the various theory-elements of **Cl-2.2**. First, it is to be noticed that its conceptual framework is a conceptual extension of **Cl-1** since the models have to contain now a new concept, T. Formally, this means that the models of **Cl-1** reappear now as *substructures* of the models of **Cl-2.2**. These models are determined by the fundamental law expressing the relationship between heat and absolute temperature in all kinds of cycles.

Cl-2.2: $x \in M_0[\text{Cl-2.2}]$ iff: $\exists K, Z, P, V, t, Q, W, A, T, f^Q, f^W, f^T$, such that

(0) $x = \langle K, Z, \Re, A, P, V, t, Q, W, T \rangle$
(1) Z is a cycle
(2) $Q = f^Q(P, V, t, \ldots)$
(3) $W = f^W(P, V, t, \ldots)$
(4) $A > 0$
(5) $Q = U + A \cdot W$
(6) $T = f^T(t)$
(7) $\forall k \in K \oint \dfrac{dQ(k, z)}{T(k, z)} \cdot dz \geq 0$

10 Specializations

A first specialization concerns reversible cycles.

REV: $x \in \mathbf{M}[\text{REV}]$ iff:

(0) $x \in M_0[\text{Cl-2.2}]$
(1) $\forall k \in K \oint \dfrac{dQ(k, z)}{T(k, z)} \cdot dz = 0$

This specialization may, in turn, be specialized in order to obtain Clapeyron's law. For this, a new theoretical concept is introduced: "Carnot's function", C. In this way, the elements of **M[REV]** become substructures of the Clapeyron models.

CLAP: $x \in \mathbf{M}[\text{CLAP}]$ iff:

(0) $x = \langle K, Z, \Re, A, P, V, t, Q, W, T, C \rangle$
(1) $\langle K, Z, \text{IR}, A, P, V, t, Q, W, T, C \rangle \in \mathbf{M}[\text{REV}]$

Fig. 4 The third theory-net
of Clausius's
thermodynamics

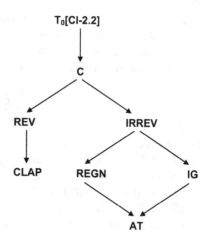

$$(2) \quad \frac{1}{T(k,z)} \cdot \frac{dT(k,z)}{dt(k,z)} = \frac{A}{C(k,z)}$$

The other specialization line concerns irreversible cycles.

IRREV: $x \in$ **M[IRREV]** iff:

(0) $x \in$ **M$_0$[Cl-2.2]**

(1) $\forall k \in K \oint \dfrac{dQ(k,z)}{T(k,z)} \cdot dz > 0$

From this, we may obtain a further specialization by admitting Regnault's additional
hypothesis above-mentioned.

REGN: $x \in$ **M[REGN]** iff:

(0) $x \in$ **IRREV**
(1) $dQ/dV = A \cdot P$

Within this same branch of the net, we may get still another specialization by simply
incorporating the ideal gas law we have already reconstructed, **M[IG]**. It is not
necessary to repeat its formulation here.

By combining **M[REGN]** with **M[IG]**, that is, by constructing a specialization
common to both specializations that we may call "**AT**" (for "absolute temperature"),
or in model-theoretic terms, by taking **M[AT]** = **M[REGN]**∩**M[IG]**, the following
result may be obtained:

If $x \in$ **M[AT]**, then: $T = 273 + t$.

Accordingly, the theory-net **Cl-2.2** has the form (Fig. 4).

We may now compare the three theory-nets implicit in Clausius's writings
examined in our exposition, **[Cl-1]**, **[Cl-2.1]**, and **[Cl-2.2]**. It is to be noted that
some (but not all) specializations of **[Cl-1]** are *embedded* into the nets **[Cl-2.1]** and
[Cl-2.2], while some of those in **[Cl-2.1]** are embedded into **[Cl-2.2]**.

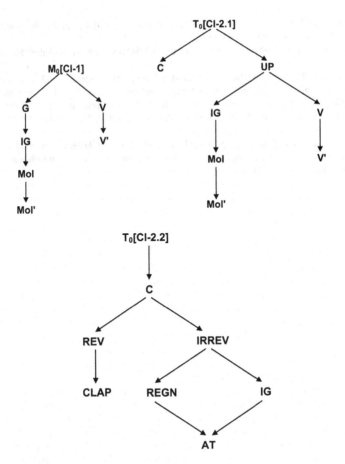

Fig. 5 The crystallization of Clausius's thermodynamics

The three theory-nets represent three consecutive phases in the crystallization process of phenomenological thermodynamics in the middle of the nineteenth century. (Fig. 5)

References

1. Balzer, W., Moulines, C.U., Sneed, J.D.: An Architectonic for Science. D. Reidel, Dordrecht (1987)
2. Caamaño, M.C.: A Structural Analysis of the Phlogiston Case. Erkenntnis **70**(3), 331–364 (2009)
3. Clausius, R.J.: Über die bewegende Kraft der Wärme. Poggendorff's Annalen der Physik **XCII** (1850)

4. Clausius, R.J.: Über eine veränderte Form des zweiten Hauptsatzes der mechanischen Wärmetheorie. Annalen der Physik **93** (1854)
5. Clausius, R.J.: Über die Art der Bewegung, die wir Wärme nennen. Annalen der Physik **100** (1857)
6. Elkana, Y.: The Discovery of the Conservation of Energy. Hutchinson, London (1974)
7. Gibbs, J.W.: On the Equilibrium of Heterogeneous Substances, Part I, Transactions of the Connecticut Academy of Arts and Sciences, vol. 3 (1876), reprinted in The Scientific Papers of J. Willard Gibbs (ed. by Bumstead, H.A., Van Name, R.G.), New York; 1906 (2nd edition, New Haven, 1957)
8. Moulines, C.U.: The Classical Spirit in J. Willard Gibbs's Classical Thermodynamics. In: Martinas, K., Ropolyi, L., Szegedi, P. (eds.) Thermodynamics: History and Philosophy. World Scientific, Singapore (1991)

Index

B.-O. Küppers et al. (eds.), *Evolution of Semantic Systems*,
DOI 10.1007/978-3-642-34997-3, © Springer-Verlag Berlin Heidelberg 2013

231

Printed in the United States
By Bookmasters